Electrochemical Studies of Biological Systems

Donald T. Sawyer, EDITOR

University of California

A symposium sponsored

by the Division of

Analytical Chemistry at

the 172nd Meeting of the

American Chemical Society

San Francisco, Calif.,

August 30, 1976

ACS SYMPOSIUM SERIES 38

AMERICAN CHEMICAL SOCIETY

WASHINGTON, D. C.   1977

Library of Congress CIP Data

Electrochemical studies of biological systems.
  (ACS symposium series; 38)

  Includes bibliographical references and index.
  ISSN 0097-6156

  1. Electrochemical analysis—Congresses. 2. Biological
chemistry—Congresses.
  I. Sawyer, Donald T. II. American Chemical Society.
Division of Analytical Chemistry. III. Series: American
Chemical Society. ACS symposium series; 38.

QD115.E524   574.1'9283   76-30831
ISBN 0-8412-0361-X   ACSMC 8   38 1–216

FOREWORD

The ACS Symposium Series was founded in 1974 to provide a medium for publishing symposia quickly in book form. The format of the Series parallels that of the continuing Advances in Chemistry Series except that in order to save time the papers are not typeset but are reproduced as they are submitted by the authors in camera-ready form. As a further means of saving time, the papers are not edited or reviewed except by the symposium chairman, who becomes editor of the book. Papers published in the ACS Symposium Series are original contributions not published elsewhere in whole or major part and include reports of research as well as reviews since symposia may embrace both types of presentation.

CONTENTS

v

PREFACE

Electrochemistry has enjoyed a renaissance during the past decade because of its use for chemical characterization. In particular, the subdisciplines of organic, inorganic, and biological chemistry have found electrochemical methods uniquely effective for determining the stoichiometries, thermodynamics, and kinetics of electron transfer reactions. Although cyclic voltammetry and controlled potential electrolysis are by far the most used electrochemical techniques, a number of new methodologies that combine electrochemical and spectroscopic measurements have been developed in recent years.

This symposium was organized in the belief that the utility of electrochemical methods for the characterization of biological systems needs to be brought to the attention of chemists and biochemists. Much of biology and biochemistry involves oxidation–reduction processes, atom-transfer reactions, and electron-transfer reactions. Because the theory and principles of electrochemistry are concerned with the same kinds of processes, as well as with the thermodynamics and kinetics of heterogeneous redox processes, substantial synergistic benefits can result from a coordinated, rational application of electrochemical principles and theories to the electron-transfer and oxidation–reduction chemistry of biology.

The twelve papers of the symposium provide a representative cross section of the kinds of electrochemical methodologies that are used to study biological systems. They also illustrate the kinds of biological problems that are being studied by such methods. Beyond cyclic voltammetry and controlled potential coulometry, the use of optically transparent thin-layer electrodes (OTTLE), rotating ring-disc enzyme electrodes, mediator titrants, differential capacitance and electrocapillary phenomena, and differential pulse polarography are discussed. The applications range from the analysis of NTA and EDTA in water samples to the characterization of the redox chemistry for several metalloproteins. Several chapters emphasize the development of improved electrochemical techniques and instrumentation for the study of biological systems. However, the major emphasis of the papers is the study of the redox properties of model compounds for biological systems. The specific systems include vitamin B_{12}, cytochrome c, cytochrome c oxidase, metal porphyrins, nitrogenase, mitochondrial superoxide dismutase, purines and pyrimidines, and a model for a mammalian heart.

The assistance of H. B. Mark, Jr., and G. S. Wilson, who chaired the two sessions of the symposium, is gratefully acknowledged. My sincere thanks to John Miller, Chairman of the Analytical Chemistry Division, for his support and encouragement in the organization of the symposium and to Marian Mann for her assistance with the correspondence and manuscript preparation.

Department of Chemistry   DONALD T. SAWYER
University of California
Riverside, Calif. 92502
November 18, 1976

Spectroelectrochemical Investigation of Vitamin B_{12} and Related Cobalamins

HARRY B. MARK, JR., and THOMAS M. KENYHERCZ
Department of Chemistry, University of Cincinnati, Cincinnati, Ohio 45221

PETER T. KISSINGER
Purdue University, W. Lafayette, Ind. 47907

This paper discusses three aspects of the spectroelectrochemical study of Vitamin B_{12} and related cobalamins: (i) the time resolved spectral study of the autooxidation of cob(I)alamins under various experimental conditions; (ii) the electrochemical behavior of the cob(III)alamins, and (iii) the electrochemical behavior of 5'-deoxyadenosylcobalamin. Several previously unknown features concerning the redox chemistry of these unusual but important complexes are reported.

The Autooxidation of B_{12}s Under Various Conditions

Recent spectroelectrochemical investigations of the oxidation of cob(I)alamins to cob(III)alamins[1] in various media has yielded the previously unobserved sequence of intermediates and steps involved in the mechanism of this biologically important reaction.[2] This sequence of intermediates disagrees with previously speculated mechanisms[2-8]. The time resolved visible-UV spectra for the reoxidation of the electroreduced Vitamin B_{12} type compounds: cyanocobalamin (B_{12}), aquocobalamin ($B_{12}a$) and dicyanocobalamin (B_{12}-CN) are reported below. The autooxidation of the cob(I)alamins were carried out under both air and inert argon atmospheres, the electrochemical reoxidation was also studied under these conditions. Mechanisms consistent with the qualitative kinetic data obtained from time resolved spectra are presented.

The recently developed mercury coated nickel minigrid system was employed in a thin layer electrolysis cell as an optically transparent electrode, Hg-Ni OTTLE.[10-13] The Hg-Ni OTTLE cell was mounted in a computerized Harrick rapid scanning dual team spectrophotometer,[14] while all experimental procedures, interface design, electrochemical instrumentation and computer programming including data acquisition, processing and reduction have been described previously.[15] All solutions (except that noted in Figure 2) were 1mM in cobalamin and 1.0M in Na_2SO_4 as the supporting electrolyte (the solutions for the B_{12}-CN experiments were also 0.1M in NaCN). The solutions were introduced into the OTTLE cell with cobalt in the +3 valence state. The solutions were then potentiostated at -1.0 volts vs. SCE until the spectra changed completely to that of the Co(I) species , (see the first spectrum of Figure 1) and remained constant (for approximately one-half hour). The cob(I)alamin solutions were allowed to undergo autooxidation in either the presence of air diffusing into the OTTLE cell or under an inert argon atmosphere. The potentiostat was disconnected during these autocatalysis experiments. During electroreoxidation step-wise (50 mv increment) potentiostatic polarization (potentiostated at each potential until spectra ceased changing) under an inert argon atmosphere was employed.

Curve A of Figure 2 shows the typical transient spectrum of a partially air reoxidized solution of cob(I)alamin while curve B is the spectrum of aquocobalamin, the final oxidation product.[9] The peaks of interest here, which correspond to intermediates in the oxidation sequence occur at 410 nm and 475 nm. During the large scale preparation of B_{12a} it was observed that the relative ratio of the 410 and 475 nm peaks varied markedly depending on the rate at which oxygen was introduced into the cob(I)alamin solution. For example, curve A of Figure 2 is the spectra obtained on bubbling oxygen through a rapidly stirred B_{12s} solution (the 410 nm peak is much larger than the 475 nm peak). However, the 475 nm peaks in the curves shown in Figure 3 are larger than the 410 nm peaks. In the latter case, air diffuses into the B_{12s} solution slowly from the edges of the OTTLE cell. Thus it is felt that the 410 and 475 nm peaks represent two different species even though previous workers have reported both peaks as being characteristic of the so-called B_{12r}!

The spectra obtained following the exhaustive reduction of B_{12}, B_{12a}, and B_{12}-CN at -1.0 volts vs. SCE were identical cob-(I)alamin species corresponding to those previously designated B_{12s}. Though the exact coordination geometry for cob(I)alamin is unknown, it has been suggested that the benzimidazole is in a base-off configuration[17] with water molecules occupying each of the axial positions.

As B_{12} and B_{12a} appear to reoxidize at comparable rates and have similar time resolved spectral characteristics, all arguments made for B_{12} are equally applicable to B_{12a}. Also, it was found that the time resolved spectral sequences and rates of peak changes were virtually the same in the presence of air or argon. The time resolved spectra for the air reoxidation of a cob(I)-alamin solution obtained by the exhaustive reduction of cyano-cob(III)alamin is shown in Figures 1 and 3. Figure 1 shows that the cob(I)alamin, as monitored by the 385 nm peak, is virtually reoxidized completely to a cob(II)alamin in the first 100 seconds. The peak which develops at 475 nm corresponds to a cob(II)ala-min,[16] (designated here as B_{12r}) grows to a maximum in the first 400 seconds and then slowly decreases finally vanishing at about

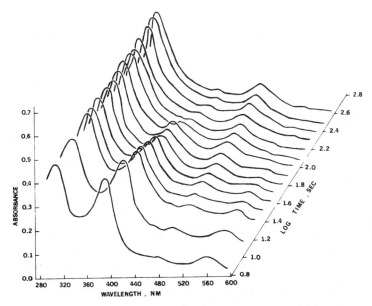

Figure 1. Time-resolved spectra for the reoxidation of cob(I)alamin to cob(II)alamin in 1.0M Na₂SO₄ at pH = 7.0 in 0–400 sec

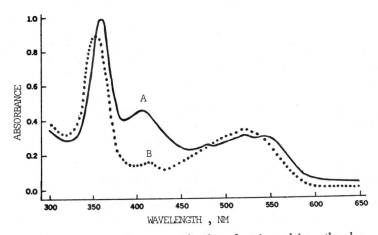

Figure 2. Spectra of the air-reoxidized product formed from the electroreduction of the cyanocob(III)alamin in 0.1M NaNO₃. A, partially reoxidized cob(II)alamin; B, totally reoxidized cob(III)alamin, aquo-cob(III)alamin.

5×10^4 seconds. The cyanocob(III)alamin peak of 360 nm which begins to develop at 10^4 seconds is completely reconverted to B_{12} by 10^5 seconds (as seen in Figure 3). The previously unreported intermediate exhibits a 410 nm peak, which occurs in the same qualitative time period of the characteristic 475 nm cob-(II)alamin peak, appears in 200 seconds, reaches a maximum by 5×10^3 seconds and has disappeared by about 2×10^4 seconds. As there are no peaks in the region of 370 nm which would identify either Co(I) or Co(III) species[16] during this time interval, it is felt that the pronounced 410 nm peak indicates a second, different cob(II)alamin intermediate which will be identified as B_{12r}'. As mentioned previously, the time resolved spectra for the air reoxidation of cob(I)alamin obtained by the exhaustive reduction of aquocob(III)alamin, is qualitatively the same as represented in Figures 1 and 3. However, the air reoxidation of cob(I)alamin obtained upon exhaustive reduction of dicyanocob(III)alamin (in the presence of 100-fold excess of CN^-) is quite different as shown in Figure 4. First of all, dicyanocob-(III)alamin is totally regenerated in less than 400 seconds. The increase/decrease in the 290 nm band, the rise and fall of the 475 nm (cob(II)alamin) peak, the final rise of the 368 nm peak plus the total lack of a 410 nm peak indicates that the reoxidation of cob(I)alamin in the presence of excess cyanide goes through only a cob(II)alamin, B_{12r} type intermediate.

Electrochemical reoxidation of cob(I)alamin, obtained by the electroreduction of B_{12} under an argon atmosphere, goes through both the B_{12r} (475 nm) and B_{12r}' (410 and 475 nm) intermediates in the potential region from -0.60 to -0.01 volts vs. SCE yielding B_{12} at +0.10 volts vs. SCE. The conditions of the electrochemical reoxidation experiments indicate that the 410 nm peak is not indicative of an oxygen adduct type of cobalamin species.

The time resolved autooxidation spectra of a cob(I)inimide[18] (no benzimidazole moiety on the corrin ring system) was also examined. The absorbance-potential reduction characteristics of the cyanoaquocob(III)inimide[20] starting material and the rate of autooxidation are parallel to those of Vitamin B_{12} itself indicating that the lack of the benzimidazole moiety has not appreciably altered the redox properties of the central cobalt ion. However, the time resolved autooxidation spectra of the cob(I)-inimide do not exhibit a 410 nm peak.

As variation of peaks in the 400-500 nm region have previously been associated with changes in the axial ligands,[19] it is attractive to speculate at the point that the two cob(II)alamins represent configurations where the benzimidazole is either coorrdinated to the Co(II), a base-on form, or where the benzimidazole has been replaced by a water in the y-axial position, a base-off form. Though the possibility exists that there is some alteration in the corrin ring structure could also account for the observed behavior, it is felt that redox changes in the ring would be irreversible. The fact that the cobinimide and

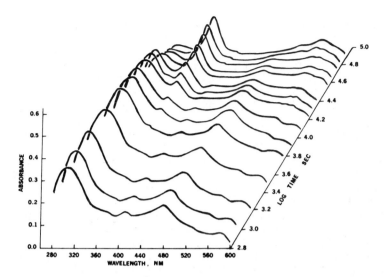

Figure 3.   Time-resolved spectra from the reoxidation of cob(II)alamin to a cob(III)alamin in 1.0M Na₂SO₄ at pH = 7.0 in 630–70,000 sec

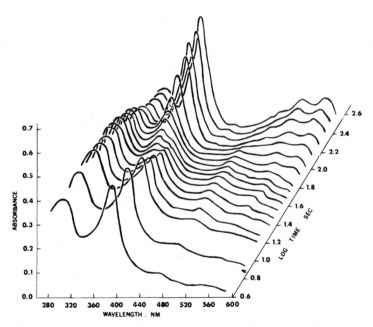

Figure 4.   Time-resolved spectra for the reoxidation of cob(I)alamin to dicyanocob(III)alamin in 1.0M Na₂SO₄ and 0.1M NaCN at pH = 11.0 in 0–410 sec

dicyanocobalamin experiments show no 410 nm band is consistent with the base-on/base-off suggestion. Furthermore, if this is correct, then the B_{12r}' would be the base-on configuration and B_{12r} would be the base-off form.

Qualitatively, the time resolved spectra indicate that the reoxidation of cob(I)alamin in the presence of stoichiometric or less amounts of CN^- follows the reaction scheme illustrated below.

As previously mentioned, the presence of excess cyanide ions affects the reoxidation scheme such that either the B_{12r}' species is not formed or that the oxidation of the B_{12r} species is kinetically favored. It has been postulated by others[3-8] that cob(I)-alamin autooxidizes to cob(II)alamin with the evolution of hydrogen and that cob(II)alamins disproportionate in the mechanism to form cob(I) and cob(III) alamin species. Under present experimental conditions this disproportionation mechanism of cob(II)-alamin is not directly observed. This is in agreement with the conclusions of Birke et al.[8] who estimated that the thermodynamic and kinetic parameters for such a disproportionation are very unfavorable. However, the time resolved spectral sequence observed does not rule out the possibility that B_{12r} undergoes disproportionation. If the rate of disproportionation of B_{12r} is very slow compared to the rate of oxidation of B_{12s} the same time resolved spectra would be obtained. It is interesting to note that we do not find any direct evidence for H_2 evolution which is easily observed (trapped small bubbles) in the OTTLE cell for systems where it occurs. We have been unable to identify the oxidizing agent(s) thus far. Furthermore, it has not been possible to calculate meaningful kinetic parameters from the time and potential resolved spectra as no quantitative diffusion model has been postulated. Therefore, calculations of diffusion in the OTTLE type cell cannot be made.

Furthermore, we cannot tell if B_{12s} oxidizes directly to both B_{12r} and B_{12r}' at different rates or that if B_{12r}' results simply from a rapid equilibrium with B_{12r} as illustrated by the dotted arrow in the proposed mechanism[11] (Cobalt +2 complex of this type are always labile.[21]) The same argument applies to the interpretation of B_{12r} and B_{12r}' oxidizing to cob(III)alamin. However, it does appear that the reoxidation of B_{12r} to dicyano-cob(III)alamin occurs much more rapidly than the reoxidation of either B_{12r} or B_{12r}' to B_{12} or B_{12a}. Diffusion studies are now in progress as well as a similar study with methyl and 5'-deoxyadenosylcob(III)alamin. Quantitative studies of the chemical and electrochemical oxidation kinetics and mechanisms will be reported at a future date.

The Electrochemical Behavior of Cob(III)alamins

The electrochemical behavior of vitamin B_{12} (cyanocob(III)-alamin) and related cobalamin compounds in aqueous media is of

importance for elucidating the biomechanistic reaction sequences
which involve cobalamin species.   There has been considerable
study of the redox processes of cobalamins using the conventional electroanalytical techniques of polarography,[9,22-33] coulometry,[3,9,34] and cyclic voltammetry,[8,35-37] and diverse working
electrode materials such as mercury[8,9,22-37] and platinum.[32,37]
However, the interpretation of the electrochemical data to unam-
biguously determine even the most fundamental parameters such as
the thermodynamic redox potentials, the number of electrons (n-
values) involved in the electron transfer steps, and the sequence
of steps in the mechanism has not been possible because of numerous complicating conditions.   The complications encompass strong
adsorption of both reactant and product, irreversibility of the
redox reactions, unusual medium effects involving the solvent
system and the supporting electrolyte, and marked variation of
electrode kinetics with the electrode material.[9,22-33]   Recently
new techniques employing minigrid electrodes in conjunction with
thin layer electrolysis cells[38] have been developed which have
proved useful to the study of the basic redox properties of cytochrome c.[9-11]   This paper reports the results obtained by using
thin layer minigrid electrode cells to study the electrochemical
and spectroelectrochemical behavior of cyanocobalamin (B$_{12}$),
aquocobalamin (B$_{12a}$), and dicyanocobalamin (B$_{12}$-CN).

<u>Spectroelectrochemistry of the Cobalamin Systems</u>.   As
changes in the valence of cobalt, the central metal ion of the
cobalamins, are reflected by distinct changes in the visible absorption spectra, a coupling of electrochemical and spectroscopic
measurements was performed to elucidate the redox behavior of the
cobalamins.   Spectroelectrochemical experiments were carried out
using the *o*ptically *t*ransparent *t*hin *l*ayer *e*lectrochemical cells
(OTTLE) in the presence and absence of the electron transfer
mediator, 2,6-dichlorophenolindophenol.[9]   The cobalamin-contain-
ing OTTLE cells were potentiostated while the optical absorbance
of a peak of interest and current levels were monitored.   When
both the absorbance stopped changing and the current levels had
fallen to essentially zero (≤ 0.1 µA), the spectrum of the solution in the OTTLE cell was recorded.   Figure 5 shows how the
spectra of a B$_{12}$ solution varies as the applied potential is
changed.   Curve 1 of Figure 5 for which the Hg-Ni minigrid electrode was potentiostated at 0.00 V is a typical spectrum for B$_{12}$
(a cob(III)alamin) with characteristic peaks at 520 and 550 nm.[40]
The spectrum obtained by potentiostating at -0.600 V (curve 2,
Figure 5) shows that the concentration of the cob(III)alamins is
decreasing as seen by the decrease in the 520- and 550-nm peaks
and the development of a new peak at 475 nm.   This 475-nm peak is
typical of that reported for B$_{12r}$, a cob(II)alamin species.[40]   On
potentiostating at -1.0 V, the spectrum obtained matches that obtained by other workers[26,40] for B$_{12s}$, a cob(I)alamin species
with a weakly absorbing broad peak at 560 nm and a tapered

shoulder in the region of 460 nm.

The quantitative change in the various peak absorbance values as a function of applied potential for B_{12}, B_{12a}, and B_{12}-CN are shown in Figures 6 through 13. Curve A of Figure 2 shows the effect of potential at a Hg-Ni minigrid electrode on the absorbance of the 550-nm peak of B_{12} as it is reduced. The B_{12} is totally reduced to a cob(II)alamin over a relatively narrow potential range (approximately 200 mV) with an absorbance "half-wave potential" of about -0.63 V. The slight increase in the absorbance between -0.8 and -1.0 V is the result of the further reduction of cob(II)alamin to cob(I)alamin which would be expected, as B_{12s}, a cob(I)alamin, exhibits a broad peak in the region of 560 nm.[40] The effect of potential on the absorbance at 550 nm for the reoxidation of cyanocob(I)alamin is shown by curve B of Figure 6. The quantitative reoxidation of the cob(I)- to cob(II)alamin occurs over the same potential range (-1.0 to -0.8 V) as shown in Figure 6 and is coincident with the behavior observed in curves A and B of Figure 7 (absorbance vs. potential curves for the 475-nm peak; the cob(II)alamin in the same potential region. However, the reoxidation of the cob(II)alamin to B_{12} occurs only when the potentials are 400 mV positive to those of the reduction potentials as can be seen from the hysteresis in curves A and B of both Figures 6 and 7 in the -0.1 to -0.7 V range. A shorter potential scan, -1.0 to -0.8 V, OTTLE experiment with the same conditions as Figure 7 (going only to the cob(II)alamin) showed the exact same hysteresis. It is important to note that B_{12} appears to be completely regenerated as the optical absorbance eventually returns to its initial value (seen in Figure 6). Figure 7, however, appears to be contradictory with respect to the reoxidation part of the above explanation. If the 475-nm peak, corresponding to cob(II)alamin formation, is to be used as an accurate indicator of cob(I)-, cob(II)-, or various cob(III)alamin species being present, then it would seem that in addition to reforming B_{12}, another cob(III)alamin may also have been formed. Spectral data obtained on the subsequent reduction of the cobalamin formed following the reoxidation of B_{12} (curve B, Figure 7), indicates a slight rise in the initial portion of the cobalamin absorbance-potential wave (monitored by the development of the 475-nm peak). The magnitude of this absorbance-potential rise remains constant as the cobalamin is recycled potentiostatically. The initial rise does not become an appreciable portion of the B_{12} absorbance-potential curve during the potentiostatic cycling process, but does resemble the behavior of B_{12a} shown in curve A of Figure 11.

The same set of experiments was performed for B_{12}-CN at a Hg-Ni electrode. As can be seen from Figures 8 and 9 dicyanocobalamin behaves quite similarly to B_{12}, the only difference being the potential region where B_{12}-CN reoxidation occrs. The hysteresis in the dicyanocobalamin reoxidation is significantly less than for B_{12} (only about 180 mV difference in the half-absorbance

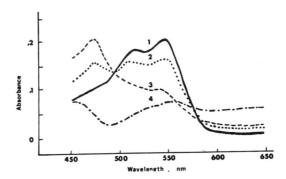

Figure 5. Spectropotentiostatic curves for the reduction of B_{12} at various applied potentials vs. SCE in a Hg–Ni OTTLE. 1, B_{12} potentiostated at 0.000 V vs. SCE; 2, B_{12} potentiostated at −0.600 V vs. SCE; 3, B_{12} potentiostated at −0.660 V vs. SCE; 4, B_{12} potentiostated at −1.000 V vs. SCE.

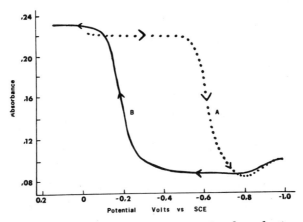

Figure 6. Potential-absorbance curves for the reduction (A) and oxidation (B) of 1.2 mM B_{12} monitored at 550 nm. 1.0 M Na_2SO_4; 0.1 M $NaNO_3$; pH 7.0; Hg–Ni minigrid; cell thickness, 0.017 cm.

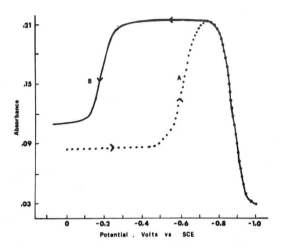

Figure 7. Potential-absorbance curves for the reduction (A) and oxidation (B) of 1.2 mM B_{12} monitored at 475 nm. 1.0 M Na_2SO_4; 0.1 M $NaNO_3$; pH 7.0; Hg–Ni minigrid; cell thickness, 0.017 cm.

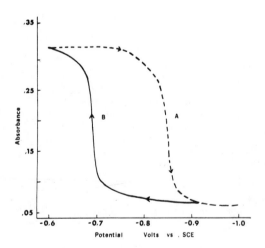

Figure 8. Potential-absorbance curves for the reduction (A) and oxidation (B) of 1.2 mM B_{12}-CN monitored at 580 nm. 1.0 M Na_2SO_4; 0.1 M KCN; pH 10.4; Hg–Ni minigrid; cell thickness, 0.017 cm.

potentials for B_{12}-CN reduction and reoxidation curves in Figures 8 and 9).

Vitamin B_{12a} was also investigated at a Hg-Ni electrode in a similar set of experiments. Curves A of Figure 10 (the 530-nm peak) and Figure 11 (the 475-nm peak) show that B_{12a} undergoes an unusual two-step process, as illustrated by the breaks at about -0.06 and -0.65 V, before complete conversion to a cob(II)-alamin species. Curve A of Figures 10 and 11 shows that the cob-(II)alamin is then reduced to cob(I)alamin as the potential increases from -0.8 to -1.0 V. On reoxidation, curve B of Figure 11, the cob(I)alamin is *reversibly reoxidized* to cob(II)alamin over the same potential range as in the *negative scan*. However, the B curves in both Figures 10 and 11 indicate that the reoxidation which corresponds to a quantitative regeneration of B_{12a} from the cob(II)alamin species is a single step process which oddly occurs at a potential *negative* to the first reduction step (ca. -50 mV).[41]

To check the unique spectroelectrochemical properties of B_{12a}, other samples of B_{12a} from different sources and preparations were examined, and, also, the electrochemical preparation was recycled a number of times. The spectroelectrochemical behavior at a particular wavelength for B_{12a} from the various preparations gave spectropotentiostatic curves(OTTLEgrams) identical with those presented herein. Also, spectropotentiostatic cycling of B_{12a} gave reproducible sets of curves. It is interesting that the absorbance-potential waves for B_{12a} in the OTTLE experiments do not correspond to any peaks in the cyclic voltammogram of B_{12a} at the same electrode. However, the three absorbance-potential "waves" for the reduction of B_{12a} do correlate reasonably well with the three waves observed in the previously reported polarography of B_{12a}.[9,26,27] To understand the unusual two-step process in the reduction of B_{12a} to a cob(II)alamin species and to determine if the electrode itself is playing a role in the electron transfer kinetics, the mediator 2,6-dichlorophenolindophenol was used in conjunction with the Au minigrid electrode.[10] The mediator functions as the primary electron transfer agent between the electrode and a redox system that has very slow heterogeneous electron transfer rates. Thus, the mediator accelerates the overall electrochemical reaction of the system of interest. The choice of this mediator was determined by the potential region of interest in this case (+0.2 to -0.2 V vs. SCE). The Au minigrid electrode was used to eliminate the possibility of oxidation of the working electrode material in this potential region and because the cyclic voltammograms of B_{12a} exhibited a more well-defined wave at an intermediate potential, and B_{12a} appeared to be less strongly adsorbed on the Au electrode. The absorbance changes of the 525-nm (B_{12a}) and 475-nm (B_{12r}) bands as a function of the applied potential are shown in Figures 12 and 13 respectively. Curve A of Figure 12 shows only one "wave" with a half-absorbance potential of -0.15 V in the +0.2 to -0.6 V

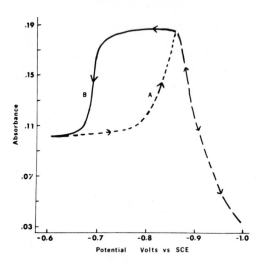

Figure 9.   Potential-absorbance curves for the reduction (A) and oxidation (B) of 1.2 mM B_{12}-CN monitored at 475 nm. 1.0 M Na_2SO_4; 0.1 M KCN; pH 10.4; Hg–Ni minigrid; cell thickness, 0.017 cm.

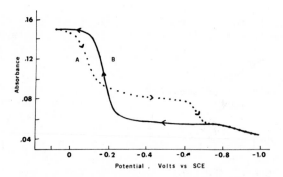

Figure 10.   Potential-absorbance curves for the reduction (A) and oxidation (B) of 0.9 mM B_{12a} monitored at 530 nm; 1.0 M Na_2SO_4; 0.1 M $NaNO_3$; pH 7.01; Hg–Ni minigrid; cell thickness, 0.017 cm.

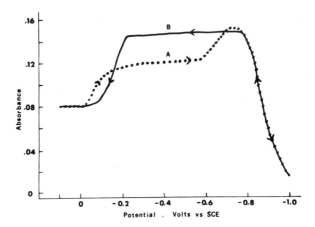

Figure 11. Potential-absorbance curves for the reduction (A) and oxidation (B) of 0.9 mM B_{12a} monitored at 475 nm. 1.0 M Na_2SO_4; 0.1 M $NaNO_3$; pH 7.0; Hg–Ni minigrid; cell thickness, 0.017 cm.

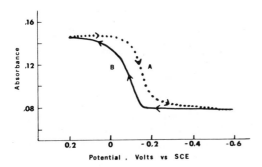

Figure 12. Potential-absorbance curves for the reduction (A) and oxidation (B) of 0.9 mM B_{12a} monitored at 525 nm. 1.0 M Na_2SO_4; 0.1 M $NaNO_3$; pH 7.0; Au minigrid and 2,6-dichlorophenolindophenol; cell thickness, 0.021 cm.

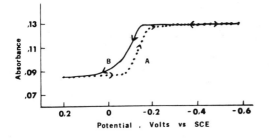

Figure 13. Potential-absorbance curves for the reduction (A) and oxidation (B) of 0.9 mM B_{12a} monitored at 475 nm. 1.0 M Na_2SO_4; 0.1 M $NaNO_3$; pH 7.0; Au minigrid and 2,6-dichlorophenolindophenol; cell thickness, 0.021 cm.

potential region scanned. From curve B of Figure 12 it can be seen that the produced B_{12r} is totally reoxidized to B_{12a} with little hysteresis (half-absorbance potential of about -0.09 V for the reoxidation) in the process. The changes in the 475-nm absorbance peak (Figure 13) again indicate only one "wave" for the generation and subsequent reoxidation of the B_{12r} with half-absorbance potentials which correspond favorably to those of the B_{12a} "waves" in Figure 12. Thus, the mediator-Au electrode system reflects a more typical redox behavior as it eliminates the unusual hysteresis effect where the reoxidation of B_{12a} from B_{12r} occurred at potentials negative to the initial reduction process (see Figures 10 and 11). However, an examination of the magnitude of the absorbance change of both the 525- and 475-nm peaks shows that it is exactly the same as that for the first absorbance waves for the Hg-Ni electrode — no mediator system (see Figures 10 and 11), indicating that even with the mediator the B_{12a} is only partially reduced at the low negative potentials. The total spectrum of the solution potentiostated at -0.6 V also indicates that part of the B_{12a} (approximately 35%) is unreacted. The same result was also obtained from the n-value studies (Table III) at both the Hg-Ni and Au minigrid electrodes. Thus, the unusual two potential processes necessary to totally reduce B_{12a} appear to be independent of both working electrode material and mediator participation. Neither the spectra for B_{12} or B_{12}-CN showed any significant reduction employing Au minigred-mediator system. No satisfactory mediator with the necessary optical and potential characteristics to explore the -0.6 to -1.0 V potential absorbance behavior at a Hg-Ni minigrid electrode has been found to date.

The half-absorbance potentials for the cobalamin species illustrated in Figures 6 through 13 are presented in Table I.

n-Value Determination. Controlled potential coulometry with a thin layer minigrid electrode system[10,11] was used to determine the number of electrons (n-value) for various waves found in the cyclic voltammograms of each of the cobalamins.[12] A typical charge vs. time curve for B_{12} is shown in Figure 14. It was necessary to extrapolate the final sloping portion of the Q-t curve back to $t = 0$ to correct for edge effects inherent in the thin layer cell system.[42] The method for correction and calculation of n-values for charging and residual current by repeating the experiment on the supporting electrolyte has been described previously.[11] The n-values, as well as the initial and final values of the applied potential steps, are shown in Table II. For the three cob(III)alamin systems using the Hg-Ni minigrid electrode, only one reduction wave is observed in the -1.0 V vs. SCE potential region and the n-value obtained in each case from the Q vs. t data was effectively two (2) yielding a cob(I)-alamin product in each case which confirms polarographic and other previously reported results[8,9,22-37] As expected, the

Table I. Half-Absorbance Potentials[a]

Working electrodes system (OTTLE)	Cobalamin species	Reduction mV vs. SCE	Oxidation mV vs. SCE	Monitored wavelength (nm)
Hg–Ni	B_{12}[b,d]	−625 (−875)	−180 (−880)	550
Hg–Ni	B_{12}[b,d]	−625 −875	−185 −875	475
Hg–Ni	B_{12}-CN[b,c]	−850	−690	580
Hg–Ni	B_{12}-CN[b,c]	−825 −910	−689 −910	475
Hg–Ni	B_{12a}[b,d]	−60 −635 (−825)	−188 (−825)	530
Hg–Ni	B_{12a}[b,d]	−75 −634 −880	−176 −878	475
Au + mediator	B_{12}[b,d]	---	---	
Au + mediator	B_{12}-CN[b,c]	---	---	
Au + mediator	B_{12a}[b,d]	−155	−93	525
Au + mediator	B_{12a}[b,d]	−140	−110	475

[a] The cobalamin concentration is 1 mM. It should be pointed out that no relationship between the half-absorbance potentials and the reversible potentials for these species exists at this time.
[b] Supporting electrolyte = 1.0 M Na_2SO_4. [c] Supporting electro-

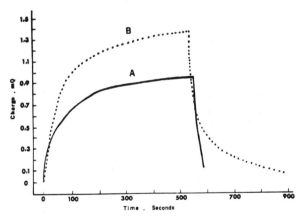

Figure 14. Charge–time curve for the application of a potential step from 0.000 to −0.970 to +0.100 V vs. SCE at a Hg–Ni OTTLE. (A) background, 1.0 M Na_2SO_4, 0.1 M $NaNO_3$. (B) B_{12}, 0.6 mM B_{12}, 1.0 M Na_2SO_4, 0.1 M $NaNO_3$.

background breakdown potential shifts positive on the Au minigrid electrode and overlaps the Co(III)-Co(I) wave observed on
the Hg-Ni electrode. On Au electrodes, B_{12} and B_{12}-CN exhibits
a small prewave at -0.3 V and a broad irreversible appearing wave
at about -0.8 V. The B_{12a} exhibits a single very broad poorly
defined wave with a peak potential at about -0.5 V. Potential
step experiments with B_{12}-CN gave fractional n-values regardless
of the magnitude of the first applied potential, the meaning of
which could not be interpreted from the electrochemical data.
The n-value for B_{12a} on a potential step to -0.6 V also yielded
a fractional value of about 0.65. It has been previously reported by some workers that only B_{12a} can be coulometrically reduced to B_{12r} (cob(II)alamin) in a one-electron step at a mercury electrode at intermediate potentials[26,31] At a Hg-Ni minigrid, the three cob(III)alamins were coulometrically reduced to
cob(I)alamin at -0.97 V and the potential was stepped to +0.1 V
and the Q vs. t curves were recorded. Only in the B_{12} case was
a reoxidation n-value equal to 2 found which indicates a virtually quantitative reoxidation to B_{12}. Fractional n-values obtained for B_{12a} and B_{12}-CN derived cob(I)alamins indicates that
only part of these cob(III)alamins are regeneraged even at
positive potentials. However, these reoxidation n-values are
difficult to interpret as complicating effects arise from the
interfering mercury(II) cyanide species which form in some cases.
At the Au minigrid electrodes only part of the cob(III)alamins
are reduced as explained above; however, it appears from the reoxidation n-values that the fraction reduced is quantitatively
regenerated at positive potentials. The ability of the base-off
cobalamin to form complexes with metal ions also obscures the
issue.[43]

Further n-value information was obtained by fixed wavelength optical monitoring techniques coupled with controlled
potential coulometry to determine n-values for appropriate redox
processes involving vitamin B_{12}. As mentioned previously B_{12}
was chosen for this investigation as earlier studies had suggested that B_{12} underwent only a single two-electron reduction
step.[9,22-33] The monitoring wavelength of 475 nm was chosen as
this peak is indicative of the presence (or absence) of a cob-
(II)alamin. Monitoring this wavelength, while coulometrically
the number of electrons transferred to the cobalamin in the process is measured, yields the n-value for each step of the mechanism. Table III summarizes the results of this spectroelectrochemical study. It is evident from the growth and decay of the
475-nm peak that a one-electron reduction does occur at intermediate potentials and that this species can undergo a further
one-electron transfer to form cob(I)alamin. The n-value in this
case cannot be determined directly because of interference from
background. This cob(I)alamin is readily reoxidized to a cob-
(II)alamin; n value equals one. Existing experimental conditions
again did not allow for an accurate determination of the n-value

Table II. *n*-Value Results

Minigrid working electrode system	Potential step region mV vs. SCE	Species	No. of electrons = n
	Reduction Coulometry		
Hg–Ni	0 to −970	B$_{12}$[a,c]	1.90
		B$_{12}$-CN[a,b]	2.00
		B$_{12a}$[a,c]	1.96
Au	0 to −500	B$_{12}$[a,c]	0.013
	+300 to −400	B$_{12}$-CN[a,b]	0.126
	+100 to −1000	B$_{12}$-CN[a,b]	1.37
	+300 to −600	B$_{12a}$[a,c]	0.65
	Oxidation Coulometry		
Hg–Ni	−970 to +100	B$_{12}$[a,c]	2.14
		B$_{12}$-CN[a,b]	0.51
		B$_{12a}$	0.38
Au	−500 to 0	B$_{12}$[a,c]	0.013
	−400 to +300	B$_{12}$-CN[a,b]	0.125
	−1000 to +100	B$_{12}$-CN[a,b]	1.40
	−600 to +300	B$_{12a}$[a,c]	0.65

[a] Supporting electrolyte = 1.0 M Na$_2$SO$_4$. [b] Supporting electrolyte = 0.1 M KCN. [c] Supporting electrolyte = 0.1 M NaNO$_3$. The cobalamin concentration is 1 mM.

Table III. Spectropotential Step *n*-Values for B$_{12}$

Working electrode system (OTTLE)	Potential step V vs. SCE From	To	Monitored wavelength (nm)	No. of electrons
Hg–Ni	Rest	−0.755	475	0.98
	−0.755	0.200	475	*a*
	Rest	−0.755	475	0.99
	−0.755	−1.000	475	0.93
	−1.000	−0.755	475	1.04
	−0.755	0.200	475	*a*

[a] Catalytic process, $n > 2$.

for the reoxidation to a cob(III)alamin.

    Conclusions. The results described above show that, in spite of the fact that the electrokinetic data are vary compli- cated, unusual, and virtually impossible to interpret mechanis- tically, the optical monitoring of the solution composition using the OTTLE technique gives a good picture of the net or overall redox reactions that take place.

    The first observation of significance is that all three cob(III)alamins (B_{12}, B_{12}-CN, and B_{12a}) undergo a quantitative one-electron reduction to either the same or similar cob(II)ala- min (B_{12r}) species at intermediate potentials in the 0.0 to -0.8 V range. Previous electrochemical studies by other groups had claimed that only B_{12a} could be reduced to B_{12r} at intermediate potentials.[6,26-31] As the polarographic and cyclic voltammetric studies did not indicate any discernible waves in this potential range for B_{12} or B_{12}-CN, it appears that no one had, therefore, attempted coulometric reductions at such potentials. However, the OTTLE results clearly show that the one-electron reaction is common to all the species but that in the case of B_{12} and B_{12}-CN the kinetics of the reaction is unusually slow even with respect to the slow scan rates employed in polarography and the cyclic voltammetry reported here. These one-electron processes for B_{12} and B_{12}-CN show up only during point-by-point potentiostatic OTTLE techniques. The reason for the extremely slow kinetics of this one-electron reaction has not been elucidated at this time. The electron transfer rate is fast enough for waves to be ob- served polarographically or with cyclic voltammetry only in the B a case. Under the same conditions the further reduction of all the cobalamin systems from the Co(II) to Co(I) oxidation state was quantitative and "reversible". The apparent hysteresis involving Co(II) to Co(III) cobalamins is not presently well un- derstood but may result from chemical reactions involved in the mechanism.

    It is interesting to note that B_{12}-CN is totally re-formed (shown in curve B, Figure 9) while cyanocob(I)alamin does not completely reoxidize to B_{12}. This suggests that B_{12} and B_{12}-CN may reoxidize by separate pathways. Because of the magnitude of the irreversibility of the B_{12} redox couple and also the fact that B_{12} is not totally re-formed (some B_{12a} appears to be a minor reoxidation product), it is thought that on electrochemical reoxidation that B_{12a} is the initial product formed and that B_{12} subsequently forms on a ligand exchange reaction involving the cyanide in solution (initially released into the solution phase during the reduction of B_{12} to B_{12r}, as shown by the fact that a -0.1 to -0.8 V OTTLE experiment (cob(III)alamin $\leftrightarrow$ cob(II)alamin) with vitamin B_{12} shows the same large irreversibility indicating that the CN^- is lost in the first reduction step). This ligand exchange reaction of B_{12a} with CN^- has been shown to be very fast.[44] However, the net rate is slow because of the dilute solutions employed. The complete regeneration of B_{12} is not possible as some CN^- is lost, probably through the formation of stable mercury(II) cyanide complexes. It was noted that the percent recovery increased on addition of excess cyanide which is consistent with this interpretation. With respect to B_{12}-CN, the final product is formed directly upon reoxidation or the follow-up ligand exchange reaction between the concentrated cyanide solution and the B_{12a}, formed by the loss of one electron from B_{12r} (with water molecules in the axial positions[45]), is very fast.

At this time it is impossible to distinguish between these two mechanisms for the reoxidation of B_{12}-CN. However, it should be noted that the B_{12r} spectra (as indicated by the 475-nm peak) in both B_{12} and B_{12}-CN reactions are virtually identical.

Perhaps the most unusual and difficult to understand result is the observation of two different potential-absorbance "waves" for the reduction of B_{12a} to a cob(II)alamin. The OTTLE spectra and the apparent n-value data indicate that B_{12a} converts to B_{12r} (about 65%) at potentials around -0.05 V at both the Hg-Ni and Au electrodes while it is necessary to raise the potential to greater than -0.6 V where the second wave corresponding to the reduction of the remaining 35% of the B_{12a} is observed. The most obvious conclusion that fits the data qualitatively is that the B_{12a} employed in these experiments was impure and contained about 35% of B_{12} itself (B_{12a} was prepared from B_{12}).   However, as pointed out above, we found that <u>all</u> batches of B_{12a} gave the same results which again would not be expected to remain constant if the various synthesis routes yielded only partial conversion. Also the spectral and polarographic properties do not suggest that any appreciable concentration of B_{12} remain unconverted and also are identical with the spectra and polarographic properties of vitamin B_{12a} produced by the totally different procedures.

Furthermore, there is considerable other indirect evidence that there is no significant unconverted B_{12} in the B_{12a} samples. Note first of all that there is no-0.6 V polarographic wave for B_{12} that corresponds to the wave for this second B_{12a} species. (It is interesting to note that previous polarographic studies had referred to the wave at -0.6 V as an impurity.)[31,46] Although the cyclic voltammogram for B_{12a} does exhibit an anodic peak at -0.28 V which could be indicative of Hg oxidation in the presence of a complexing ligand, this wave is about 50 mV positive to the peak corresponding to mercury-cyanide formation in the B_{12} cyclic voltammogram and there is no corresponding cathodic sweeps of B_{12} itself. Finally high pressure liquid chromatography (using a mixture of either 80% isopropyl alcohol and 20% water, or 65% methanol and 35% water, at 2000 psi on an Aminex A-4 column, with detector wavelength set at λ 360 nm) on B_{12a} has exhibited two closely spaced yet distinct peaks both with retention times that are different than B_{12}. Also, a thin layer chromatographic comparison of B_{12} and B_{12a} using a 65%

methanol and 35% water solvent system has shown that B_{12} and
B_{12a} have relative fronts, though no separation of B_{12a} itself
was observed. It is felt that ring structure differences would
not account for the two unique B_{12a} species, as the B_{12a} prepared
from three techniques (potentiostatic, biological and chemical)
would not give identical 65/35 ratio of concentrations. Moreover, it is hard to understand how two different rings, which
would be expected to be common to all cobalamins, exhibit drastic
reduction potential differences for B_{12a} and not for B_{12} or B_{12}-
CN. Thus, it is attractive to speculate that the two species represent differences in axial ligand configuration. The simplest
answer would be that one of the B_{12a} species contains water molecules in the X and Y positions (the "base-off" form) while the
other is in the configuration with one water in the X position
and the 5,6-dimethylbenzimidazole in the Y position (the "base-
on" form). The spectroelectrochemical data clearly demonstrate
that the two B_{12a} species are not in equilibrium. However,
Thusius has shown that the X position of B_{12a} is very labile
(rate constants of about 170-2300 M·s^{-1}).[44] However, no measurements have been made on the Y position benzimidazole-H_2O exchange
rates.[47] It is possible that this exchange could be very slow.
The fact that the diaquocob(III)inamide (having no benzimidazole
attached to the corrin ring side chain) has been reported to be
difficult to reduce ($E_{1/2} \simeq$ -0.7 V)[47] is consistent but not proof
of the "base-on"-"base-off" explanation. This fact suggests that
the "base-on" aquocob(III)alamin form has a configuration favor-
able to reduction (the -0.15 V wave) and the "base-off" form
which would closely correspond to a diaquocob(III)inamide configuration is difficult to reduct (-0.6 V wave).[47] However, recent
spectroelectrochemical studies by Lexa and Saveant[48] have shown
that at platinum grid electrodes B_{12a} does not exhibit this two
wave one-electron behavior. Only one wave is observed at about
0.0 V. Furthermore, they have also shown that the diaquocob-
(III)inamide oxidizes mercury metal spontaneously and forms the
cob(II)inamide. Thus, the reported values for this compound[47]
are incorrect and really correspond to the subsequent electrochemical reduction of the Cob(II)inamide. Thus, it appears that
the mercury electrode is in some way the cause of this two wave
one-electron observation.

It is obvious that there are many unanswered questions concerning rates of the microscopic precesses involved in the redox
chemistry of cobalamin complexes. However, the macroscopic resultant effect of electrode potential in solution composition is
now well defined. With this basic overall mechanistic information, a more comprehensive study of the electrode kinetics and
time resolved spectral studies on potential step experiments on
these and other cobalamins under variable conditions of pH, supporting electrolyte, and electrode material may elucidate all the
steps in the overall mechanism.

The Electrochemical Behavior of 5'-Deoxyadenosyl-Cobalamin (Coenzyme B_{12})

The understanding of the redox processes of the cobalamin coenzyme, 5'-deoxyadenosylcobalamin (coenzyme B_{12} or Aden-B_{12}), is fundamental in constructing a valid reaction sequence for cobalamin compounds in nature.[49,50] Of particular importance is the oxidation state of cobalt in Aden-B_{12} and its influence on the reactivity of the cobalamin species. As a matter of notational ease and to maintain consistency with existing vitamin B_{12} concepts, the cobalamin coenzyme has been primarily interpreted in terms of a Co^{3+} and an adensyl carbanion.[49] Redox mesomers consisting of a Co^{2+} and an adenosyl radical or a Co^{1+} and an adenosyl carbonium ion have also been formally considered. Although numerous publications have assumed the cobalt to be in the +3 oxidation state,[49-63] little evidence for this assignment is available. Some evidence exists which supports the claim that the cobalt of the cobalamin should be considered as a divalent species.[64-71]

Similar electrochemical and spectroelectrochemical techniques utilizing an amalgamated gold minigrid electrode in a thin layer configuration have been employed to examine the redox sequence of 5'-deoxyadenosylcobalamin. This investigation suggests that the cobalamin coenzyme undergoes a single electron reduction to form vitamin B_{12s}, indicating cleavage of the cobalt-carbon bond. The reoxidation of the reduced system containing the B_{12s} occurs via two consecutive single electron transfers resulting in the quantitative formation of vitamin B_{12a}, aquocobalamin. High performance liquid chromatography confirmed that 5'-deoxyadenosine is the ultimate form of the cleaved 5'-deoxyadenosyl moiety.

These results indicate that 5'-deoxyadenosyl-cobalamin is reduced by a single electron to form the cob(I)alamin, B_{12s}, which is the common reduction product of all cobalamin species.[12] The formation of the cob(I)alamin, B_{12s}, from Aden-B_{12} is significant as numerous publications[72] have alluded to the fact that due to the high nucleophilicity of the cob(I)alamin species it may be the biologically active form of the coenzyme. The result reported herein is the first electrochemical evidence for the formation of the cob(I)alamin from the cobalamin coenzyme. That B_{12s} can be completely reoxidized to B_{12a}, a cob(III)alamin, is significant in that existing cyclic biochemical mechanisms involve Aden-B_{12} as the initially reactive species.[72] Furthermore, ribonucleotide reductase has shown specific activity toward B_{12s} and Aden-B_{12} in the presence of 5'-deoxyadenosine.[73,74]

Acknowledgments

This research was supported in part by the National Science Foundation, NSF CHE76-04321 and the National Institutes of Health, GM-22713-01.

Literature Cited

1. Pratt, J. M., "Inorganic Chemistry of Vitamin B_{12}," Academic Press, New York, N. Y., 1972, pp. 15-17.
2. Huennekens, F. M., in "Biological Oxidations," Thomas P. Singer, Ed., Interscience Publishers, New York, N. Y., pp. 482-502.
3. Tackett, S. L., Collat, J. W., and Abbot, J. C., Biochemistry, (1963), 2, 919.
4. Collat, J. W. and Abbot, J. C., J. Amer. Chem. Soc., (1964), 86, 2308.
5. Schrauzer, G. N., Deutsch, E. and Windgassen, R. J., J. Amer. Chem. Soc., (1968), 90, 2441.
6. Yamada, R., Shimizu, S. and Fukui, S., Biochemistry, (1968), 2, (7), 1713.
7. Rudiger, H., Eur. J. Biochem., (1971), 21, 264.
8. Birke, R. L., Brydon, G. A., and Boyle, M. F., Electroanal. Chem., (1974), 52, 237.
9. Kenyhercz, T. M. and Mark, Jr., H. B., Anal. Lett., (1974), 7, 1.
10. Heineman, W. R., Norris, B. J. and Goelz, J. Anal. Chem., (1975), 47, 79.
11. Heineman, W. R., DeAngelis, T. P. and Goelz, J., Anal. Chem., (1975), 47, 1364
12. Kenyhercz, T. M., DeAngelis, T. P., Norris, B. J., Heineman, W. R. and Mark, Jr., H. B., J. Amer. Chem. Soc., (1975), 98, 2469.
13. However, experiments were performed with the minigrid electrode area increased to occupy the entire cell volume to determine if the edge effects from the diffusion of unreduced cob(III)alamins from solution not in immediate contact with the minigrid affected the time resolved spectra. No significant difference was observed.
14. Strojek, J. W., Gruver, G. and Kuwana, T., Anal. Chem., (1969), 41, 481.
15. Mark, Jr., H. B., Wilson, R. M., Miller, T. L., Atkinson, T. V., Yacynych, A. M., and Woods, H., "The On-Line Computer in New Problems in Spectroscopy: Applications to Rapid Scanning Spectroelectrochemical Experiments and Time Resolved Phosphorescence Studies" in "Information Chemistry; Computer Assisted Chemical Research Design," S. Fujiwara and H. B. Mark, Jr., Eds., University of Tokyo Press, Tokyo, Japan, 1975, pp. 3-28.
16. Beaven, G. H. and Johnson, E. A., Nature, (1955), 176, 1264.
17. Ref. 1, p. 184.
18. Ref. 1, p. 20 to 27.
19. Ref. 1, p. 55
20. The cyanoaquocol(III)inimide was prepared by a previously described method; Ref. 1, p. 294.

21. Basolo, F. and Pearson, R. G., "Mechanisms of Inorganic Reactions," John Wiley and Sons, Inc., New York, N. Y., pp. 141-144.
22. Diehl, H., Sealock, R. R., and Morrison, J., Iowa State Coll. J. Sci., (1950), $\underline{24}$, 433.
23. Diehl, H., Morrison, J. I., and R. R. Sealock, Experientia, (1951), $\underline{7}$, 60.
24. Diehl, J., and Morrison, J. I., Rec. Chem. Prog. (1972), $\underline{31}$, 15.
25. Boos, R. N., Carr, J. E., and Conn, J. B., Science, (1953), $\underline{117}$, 603.
26. Jaselskis, B. and Diehl, H., J. Am. Chem. Soc., (1959), $\underline{81}$, 4345.
27. Jaselskis, B. and Diehl, H., J. Am. Chem. Soc., (1958), $\underline{80}$, 2147.
28. Collat, J. W., and Tackett, S. L., J. Electroanal. Chem., (1962), $\underline{4}$, 59.
29. Tackett, S. L., Ph.D. Thesis, Ohio State University, 1962.
30. Kratochvil, B., and Diehl, H., Talanta, (1966), $\underline{13}$, 1013.
31. Hogenkamp, H. P. C. and Holmes, S., Biochemistry (1970), $\underline{9}$, 1888.
32. Lexa, D. and L'hoste, J. M., in "Biological Aspects of Electrochemistry," G. Milazzo, P. E. Jones, and L. Rampazzo, Ed., Birkhauser Verlag, Stuttgart, 1971, pp. 395-404.
33. Abd-el-Nabey, B. A., J. Electroanal. Chem., (1974), $\underline{53}$, 17.
34. Das, P. K. et al., Biochim. Biophys. Acta., (1967), $\underline{141}$, 644.
35. Tackett, S. L. and Ide, J. W., J. Electroanal. Chem., (1971), $\underline{30}$, 510.
36. Swetik, P. G., and Brown, D. G., J. Electroanal. Chem., (1974), $\underline{51}$, 433.
37. Kenyhercz, T. M. and Mark, Jr., H. B., in preparation.
38. Murray, R. W., Heineman, W. R., and O'Dom, G. W., Anal. Chem., (1967), $\underline{39}$, 1666.
39. Provided by Dr. E. A. Deutsch, Department of Chemistry, University of Cincinnati.
40. Beaven, G. H. and Johnson, E. A., Nature (London), (1955), $\underline{176}$, 1264.
41. It should be pointed out the reaction $B_{12a} = [H^+] + B_{12b}$ (hydroxycob(III)alamin) has a pk_a of 7.8 and a more negative reduction potential than B_{12a}: H. O. A. Hill, "Inorganic Biochemistry," Vol. 2, G. Eichcon, Ed., Elsevier, New York, N. Y., 1973, Chapter 30. This proton equilibrium is undoubtedly fast and, thus, only the reduction of the B_{12s} will be observed in these OTTLE experiments.
42. McDuffie, B., Anderson, L. B., and Reilley, C. N., Anal. Chem., (1966), $\underline{38}$, 883.
43. Cotton, F. A. and Wilkinson, G., "Advanced Inorganic Chemistry," 3d ed., Interscience, New York, N. Y., 1972, p. 519.
44. Thusius, D. J. Am. Chem. Soc., (1971), $\underline{93}$, 2629.

45. Kenyhercz, T. M., Yacynych, A. M., and Mark, Jr., H. B., Anal. Lett., (1976), 9, 203.
46. Reference 6, p. 4346.
47. Reference 11, p. 1889.
48. Lexa, D. and Saveant, J. M., University of Paris, Private Communication, 1976.
49. Babior, B. M. in "Cobalamin: Biochemistry and Pathophysiology," B. M. Babior, Ed., John Wiley and Sons, New York, N.Y., 1975, p. 141.
50. Pratt, J. M., "Inorganic Chemistry of Vitamin B_{12}," Academic Press, London, 1972, p. 296.
51. Mahler, H. R. and Cordes, E. H., "Biological Chemistry," Harper and Row, Evanston, Ill., 1971, p. 427.
52. Hill, J. A., Pratt, J. M., and Williams, R. J. P., J. Theor. Biol., (1962), 3, 423.
53. Hogenkamp, H. P. C., Barker, H. A. and Mason, H. S., Arch. Biochem. Biophys., (1963), 100, 353.
54. Hill, J. A., Pratt, J. M. and Williams, R. J. P., J. Chem. Soc., (1964), 5149.
55. Pratt, J. M., J. Chem. Soc., (1964), 5154.
56. Huennekens, F. M. in "Biological Oxidations," T. P. Singer, Ed., John Wiley and Sons, New York, N. Y., 1968, p. 483.
57. Hogenkamp, H. P. C. and Holmes, S., Biochem., (1970), 9, 1889.
58. Cotton, F. A. and Wilkinson, G., "Advanced Inorganic Chemistry," John Wiley and Sons, New York, N. Y. 1972, p. 888.
59. Hughes, M. N., "The Inorganic Chemistry of Biological Processes," John Wiley and Sons, New York, N. Y., 1974, p. 187.
60. Costa, C., Puzeddu, A. and Reisenhofer, E., Bioelectrochem. Bioenerg., (1974), 1, 29.
61. Hill, H. A. O. in "Inorganic Biochemistry," G. L. Eichhoren, Ed., Elsevier, New York, N. Y., 1975, p. 1076.
62. Babior, B. M., Acc. Chem. Res., (1975), 8, 378.
63. Abeles, R. H. and Dolphin, D., Acc. Chem. Res., (1976), 9, 114.
64. Seki, H., Shida, T., and Imamura, M., Biochem. Acta., (1974), 372, 106.
65. Hill, H. A. O., Pratt, J. M. and Williams, R. J. P., Disc. Farad. Soc., (1969), 16S.
66. Kratochvil, B. and Diehl, H., Talanta, (1966), 13, 1013.
67. Nowick, L. and Pawelkiewicz, Bull. Acad. Pol. Sci. Cl. II., (1960), 17, 433.
68. Johnson, A. W. and Shaw, N., Proc. Chem. Soc., (1960), 420.
69. Bernhauser, K., Gaiser, P., Muller, O., Muller, E., and Gunter, F., Biochem. (1961), 333, 560.
70. Johnson, A. W., Mervyn, L., Shaw, N., and Smith, E. L., J. Chem. Soc., (1963), 4146.
71. White, A., Handler, P., and Smith, E. L., "Principles of Biochemistry," McGraw Hill, St. Louis, Mo., 1973, p. 1173.
72. Schrauzer, G. N. and Sibert, J. W., J. Amer. Chem. Soc., (1970), 92, 1022, and references therein.

73. Hamilton, A., Yamada, R., Blakley, R. L., Hogenkamp, H.P.C.,
    Looney, F. D., and Winfield, M. E., Biochem. (1971), 10, 347.
74. Tamao, Y. and Blakley, R. L., Biochem., (1973), 12, 24.

2

Bioelectrochemical Modelling of Cytochrome c

CHARLES C. Y. TING and JOSEPH JORDAN

152 Davey Laboratory, Department of Chemistry, Pennsylvania State University, University Park, Penn. 16802

MAURICE GROSS

Laboratoire d'Electrochimie et Chimie-Physique du Corps Solide, Université Louis Pasteur, BP 296 R/8, 67008 Strasbroug, France

Numerous papers have in recent years been devoted to electro-chemical studies of porphyrins and metalloporphyrins. Paradoxi-cally, an investigation of heme c (the prosthetic group of cyto-chrome c) is conspicuous by its absence. Heme c was first prepared from the naturally occurring protein in a classical piece of work by Theorell (1). Subsequently both heme c and its equatorial ligand (porphyrin c) became accessible by the synthetic route (2,3) outlined in Figure 1. The salient feature of porphyrin c is the bis-cysteinated substitution on the ring, which is unique in cytochrome c among hemoproteins (4). The corresponding substi-tuents in hemoglobin and myoglobin are vinyl groups (5,6). In the present paper, we report some preliminary findings on the electro-chemical behavior of porphyrin c and heme c.

Experimental

Materials. Porphyrin c and heme c were synthesized ad hoc using procedures referred to earlier in this write-up. Yields and elemental analyses are summarized in Table I.

Table I

Compound	Mol. Wt.	% Yield	Elemental Analysis					
			% Fe		% S		% N	
			Theory	Actual	Theory	Actual	Theory	Actual
Porphyrin c	805.0	40	0	0.22	7.97	8.05	10.44	10.10
Heme c	917.8	90	6.08	6.05	6.99	6.81	9.16	8.94

Authenticity was verified with the aid of the spectra illustrated in Figure 2 recorded with the aid of a Bausch and Lomb 505 Spectro-photometer and quartz cells whose optical pathlength was 1 and 0.1 cm.

D.C. Polarography. Current-voltage curves were recorded at a conventional dropping mercury electrode (dme) which had

26

the following characteristics: m = 1.66 mg per second; t (open circuit) = 4.55 seconds.

Cyclic Voltammetry. Kemula's hanging drop mercury electrode (hdme) served as indicator electrode. Potential scan rates in a range between 0.01 and 50 volt·sec^{-1} were used.

Coulometry. Current-time integrals were determined at appropriate controlled potentials, corresponding to well-defined polarographic diffusion current domains. The cathode was a mercury pool of 2.60 sq. cm.

Instrumentation, Solvents, Supporting Electrolytes, etc. All experiments were carried out at 25°C. On solubility considerations DMF and water were used as solvents for heme c and porphyrin c respectively. 0.1 M perchloric acid was the supporting electrolyte in all experiments. Three-electrode systems were employed throughout, using a saturated aqueous calomel reference electrode (SCE) and a platinum foil auxiliary counter electrode. All electrochemical measurements were performed with the aid of a multipurpose instrument equipped with advanced solid state operational amplifier and feedback circuits, viz, the Model 170 Electrochemical System supplied by Princeton Applied Research (PAR) Corporation, Princeton, N.J. Output signals were automatically corrected for iR drops and recorded on a built-in X-Y pen recorder and/or with the aid of a digital oscilloscope (Model 1090 with Model 90 plug-in unit, Nicolet Instrument Corporation, Madison, Wisconsin). The scope was equipped with a mini-computer which had capabilities of storing data in a 4096 x 4096 array memory as fast as 5 μsec per data point. The coupling of the PAR instrument with the oscilloscope allowed measurement of fast linear sweep voltammograms with an accuracy of 1 percent.

Whenever appropriate, total currents were corrected for residual currents to yield the corresponding faradaic currents. Potentials are reported in accordance with the Stockholm Sign Convention of the International Union of Pure and Applied Chemistry (7), i.e., the more cathodic (reducing) a potential the more negative its assignment.

Results and Discussion

Experimental findings are summaried below.

Electroanalytical Chemistry of Porphyrin c.
1. D.C. polarograms of porphyrin c yielded two cathodic waves with well-defined limiting currents whose characteristics are listed in Table II.

Table II

Half-Wave Potential (volt vs. SCE)	Limiting Current		
	Potential Domain (volt)	$\int i\,dt$ (faradays·mole^{-1})	Wave Analysis* Slope (volt)
-0.525	-0.56 to -0.62	2.02 ± 0.04	0.030
-0.730	-0.88 to -1.0	3.94 ± 0.02	0.054

*Slope of plot of log $[(i_d-i)/i]$ versus E.

2. Coulometry at -0.615 volt versus SCE (on the first limiting current plateau) substantiated a two-electron transfer, i.e., two faradays per mole of porphyrin c.
3. Coulometry at -0.945 volt versus SCE (on the second limiting current plateau) indicated the occurrence of an overall four-electron transfer, i.e., a total of four faradays per mole of porphyrin c.
4. Cyclic Voltammetry at the hdme yielded the following results.
   (a) At relatively fast potential scan rates (10<v<50 volt· sec^{-1}) in a range between -0.2 and -0.7 volt Nernst-reversible cathodic and anodic peaks were obtained. The peak separation (ΔE_p) was 0.030 volt.
   (b) At slow potential scan rates in a range between -0.2 and -1.0 volt two cathodic peaks were apparent, each involving two electrons. One of these occurred at the same potential as the reversible peaks described in (a) above. However, no anodic peaks were obtained on the reverse scan.
   (c) Diagnostic plots of $i_p v^{-\frac{1}{2}}$ versus log v and of $\Delta E_p(\Delta \log v)^{-1}$ versus log v were consistent with EC-type mechanisms for both peaks.

Electroanalytical Chemistry of Heme c. Upon insertion of iron into porphyrin c, the polarographic and voltammetric results underwent a drastic change. In lieu of two discrete two-electron transfer waves, only a single quasireversible one-electron transfer was observed, both at the dropping mercury electrode by classical d.c. polarography and at the hdme by cyclic voltammetry at all scan rates. The corresponding half-wave potential was -0.340 volt which is compatible with expectations based on comparable ferri-ferroheme redox potentials (8,9).

A summary of the information which transpired is presented in Table III.

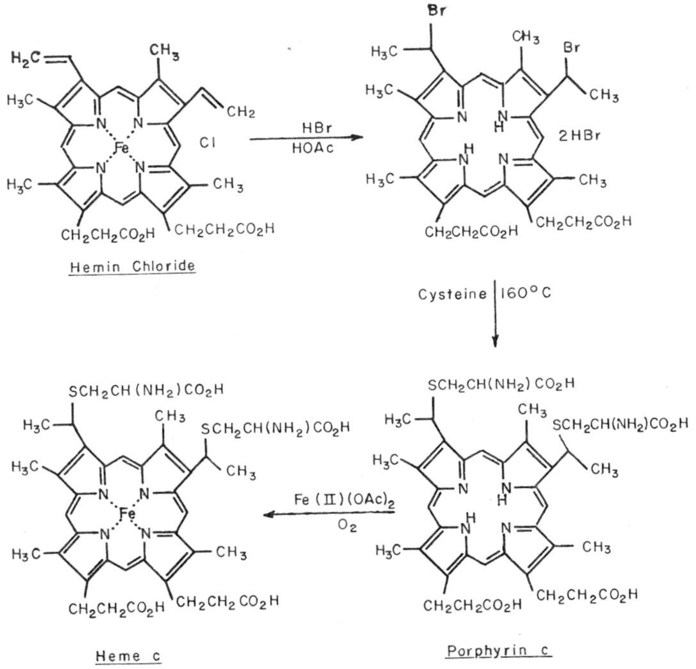

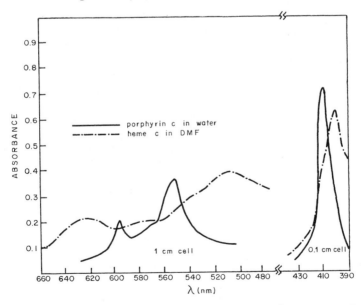

Figure 1.   Synthesis of porphyrin c and heme c

Figure 2.   Spectra of porphyrin c and heme c in the presence of
0.1 M perchloric acid

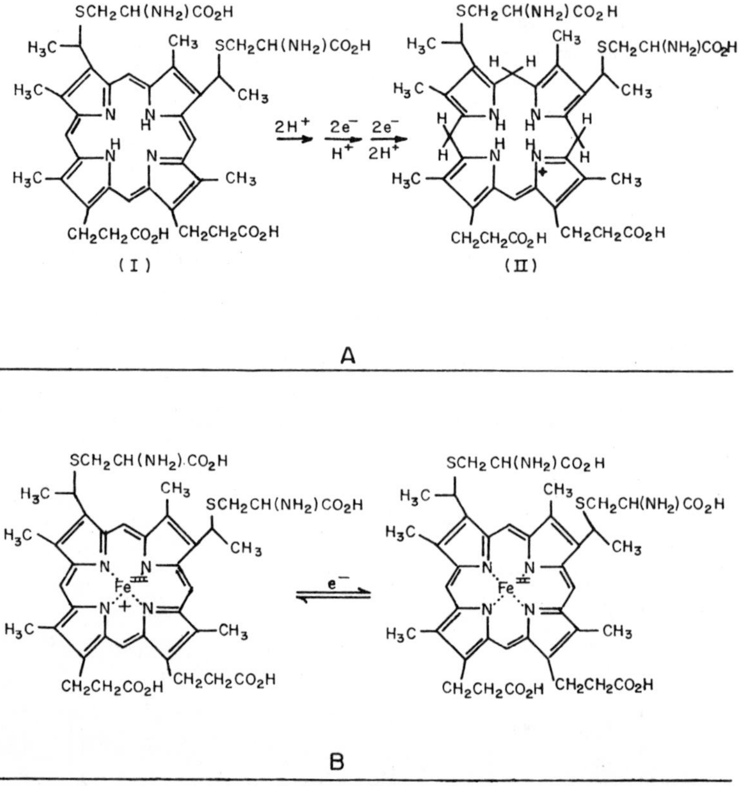

Figure 3. Electrochemical reaction mechanisms. A, conversion of por-phyrin c (I) to corresponding porphomethene moiety (II); B, electrooxida-tion–reduction of heme c.

Table III

Synopsis of the Electroanalytical Behavior of Porphyrin c and Heme c

Electroactive Moiety	Total Number of faradays per mole	Number of Discrete Electron Transfer Steps	Type of Electron Transfer
Porphyrin c	4	2 steps of 2 electrons each	EC
Heme c	1	1	Quasi-reversible

By analogy with known mechanisms previously substantiated for other porphyrins (10), we postulate that porphyrin c was electro-reduced via two successive two-electron transfer sequences, yielding a porphomethene product as shown in Figure 2, Part A. Each two-electron-transfer step was coupled with a subsequent protonation. In contradistinction, insertion of iron into the porphyrin blocked all the electron transfer sites operative in porphyrin c and converted the corresponding metalloporphyrin (heme c) into a single-electron donor-acceptor system as illustrated in Figure 3, Part B.

Acknowledgments

The work described in this communication was supported by the Centre National de la Recherche Scientifique (CNRS, France), by Research Grant HL 02342 from the National Heart, Lung and Blood Institute, National Institutes of Health, United States Public Health Service and by Research Grant RG 764 from the North Atlantic Treaty Organization (NATO).

Literature Cited

1. Falk, J.E., "Porphyrins and Metalloporphyrins", p 95, Elsevier, Amsterdam, 1964.
2. Neilands, J.B., Tuppy, H., Biochim. Biophys. Acta, (1960), 38, 351.
3. Gnichtel, H., Lautsch, W., Ber., (1965), 98, 1647.
4. Dickerson, R.E., Takano, T., Eisenberg, D., Kallai, O.B., Samson, L., Cooper, A., Margoliash, E., J. Biol. Chem., (1971), 246, 1511.
5. Perutz, M.F., Nature, (1970), 228, 726.
6. Kendrew, J.C., Dickerson, R.E., Strandberg, B.E., Hart, R.G., Davies, D.R., Phillips, D.C., Shore, V.C., Nature, (1960), 185, 422.
7. McGlashan, M.L., Pure and Appl. Chem., (1970), 21(1), 3.
8. Feinberg, B.A., Gross, M., Kadish, K.M., Marano, R.S., Pace, S.J., Jordan, J., Bioelectrochem. Bioenerg., (1974), 1, 73.
9. Betso, S.R., Klapper, M.H., Anderson, L.B., J. Am. Chem. Soc., (1972), 94, 8197.
10. Neri, B.P., Wilson, G.S., Anal. Chem. (1972), 44, 1002.

3

Control of the Potentials of Metal Ion Couples in Complexes of Macrocyclic Ligands by Ligand Structural Modifications

DARYLE H. BUSCH, DALE G. PILLSBURY, FRANK V. LOVECCHIO,
A. MARTIN TAIT, YANN HUNG, SUSAN JACKELS,
MARY C. RAKOWSKI, WAYNE P. SCHAMMEL, and L. Y. MARTIN

Evans Chemical Laboratory, The Ohio State University,
88 W. 18th Avenue, Columbus, Ohio 43210

The functions of heme-proteins in their natural systems may all be regarded as related to oxidation reduction processes. It is a characteristic feature of these substances that the actual functions they serve vary broadly with the specific heme protein and the potentials for the Fe^{2+}/Fe^{3+} couple vary correspondingly over a substantial range of values. Much of the interest in the metal complexes of synthetic macrocyclic ligands is understandably related to their structural similarity to the heme prosthetic group.

Complexes with synthetic macrocyclic ligands have provided uniquely convenient systems for evaluation of the dependence of oxidation reduction properties of metal chelates on the detailed structure of the ligand while maintaining a given metal ion in an approximately constant coordination geometry. The complexes of tetraaza tetradentate macrocycles have figured most heavily in these studies and are the subject of this report. A variety of kinds of processes are possible, depending primarily on the metal ion, the ligand and the solvent, These include oxidation and reduction of the central metal ion, various oxidation and reduction reactions of the ligand and processes which involve both the metal center and the ligand. Attention is focused here on the first category of processes.

For a large number of macrocyclic ligands, it has been possible to identify electrode processes that are attributable to the same metal ion couple. Figure 1 illustrates the range of $E_{1/2}$ values found for the Ni^{2+}/Ni^{3+} couple. The majority of the entries (vida infra) in Figure 1 involve the results of measurements in acetonitrile solutions against an Ag/Ag^+ (0.1 $\underline{M}$) reference electrode using 0.1 $\underline{M}$ (n–Bu)$_4$NBF$_4$ supporting electrolyte. In a few cases, especially the porphyrins, literature values have been adapted (1) to the same conditions by empirical corrections. The Ni^{2+}/Ni^{3+} couple has been confirmed for the processes by esr examination of the product of oxidation of the Ni^{2+} starting material in many cases (2). The most striking aspect of the data summarized

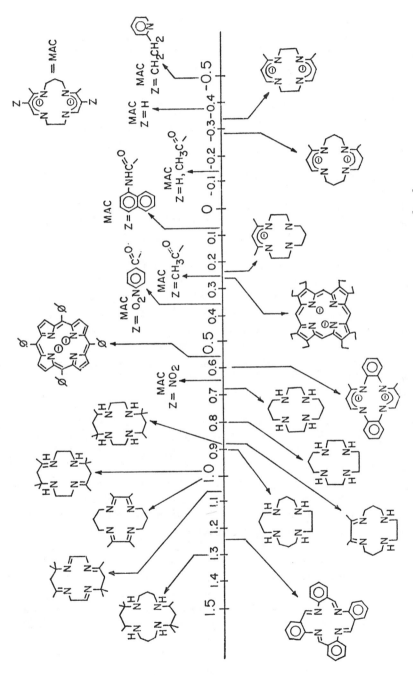

Figure 1.   E_½ for Ni^{2+}/Ni^{3+} couple vs. Ag/AgNO₃ in acetonitrile solution

in Figure 1 is the remarkable range of electrode potentials that can be realized for this single metal ion couple by variations in ligand structure. Clearly, the means are at hand to generate almost any desired value of $E_{1/2}$ for a given metal ion couple within a range approaching 2 volts. The analysis of the structure-potential relationship that follows shows in detail how this can be accomplished.

Complexes With Saturated Neutral Tetraaza Ligands. The ligands of Figure 2 all contain four secondary amine donors as members of saturated tetradentate ligands. The structures vary in ring size and in the numbers of methyl substituents on the rings. In the case of the Co^{2+}/Co^{3+} couple (3) it has been possible to examine the effect of ring size in detail using ligands 1 through 4 of Figure 2 and $E_{1/2}$ values are listed in Table I. Theoretically, these ligands can chelate in a variety of configurationally isomeric forms depending on the chiralities of the four nitrogen donors. The smaller rings, [13]aneN$_4$ and [14]aneN$_4$, produce only one structure in the complexes trans-Co(MAC)Cl$_2^+$, while [15]aneN$_4$ and [16]aneN$_4$ yield two configurational isomers (3,7). Thus, the four rings yield 6 compounds for correlation between ring size and oxidation reduction behavior. The configurations of the two isomers have been established for [15]aneN$_4$ by ^{13}C nmr combined with chemical properties and quantitative conformational analysis (3,7). These two isomers are represented by structures 1 and 2 below (isomer I of Table I has structure 1; isomer II, structure 2). This representation indicates the chirality of a nitrogen atom by the + or - sign. A plus means the hydrogen on that nitrogen is above the plane of coordination of the four nitrogens of the macrocycle; minus means it is below that plane. The numerals 2 and 3 identify the chelate rings by indicating how many carbon atoms are in each particular chelate ring. With a sophisticated computer program, the relative sizes of the metal ion sites inside the macrocyclic ligands have been estimated. The programs are based on the classic model of the strained molecule (the ligand (4,5)), and they may be used to predict the least strained conformations of the ligands with their donors appropriately arrayed for chelation (6). A very useful parameter derived from such calculations is the ideal M-N distance, a quantity that has a characteristic value for each macrocycle in each configuration. These values are also included in Table I. Since it has not been possible to definitely assign the configurations of the isomers of trans-Co([16]aneN$_4$)Cl$_2^+$ and since the $E_{1/2}$ values for the isomers are very similar only the value of ideal M-N for the most stable configuration is included in the table. Within the constraints of the model, it has been concluded that the relative values of ideal M-N are meaningful while the absolute values are probably displaced slightly from the unavailable true values. A calibration of these values has been made (5,7) using the d-d electronic spectra of the complexes and the

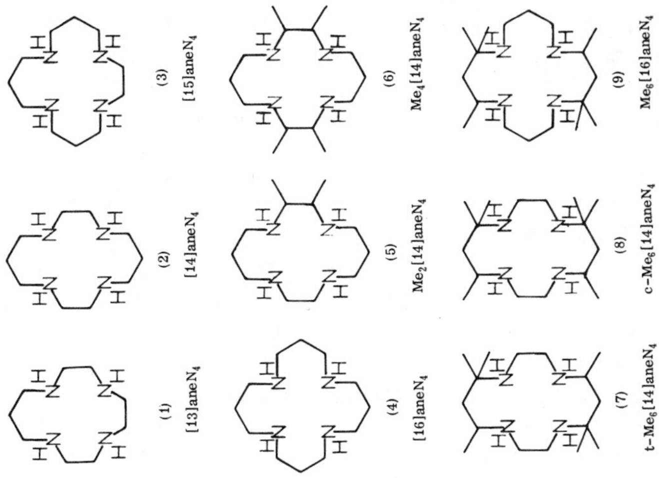

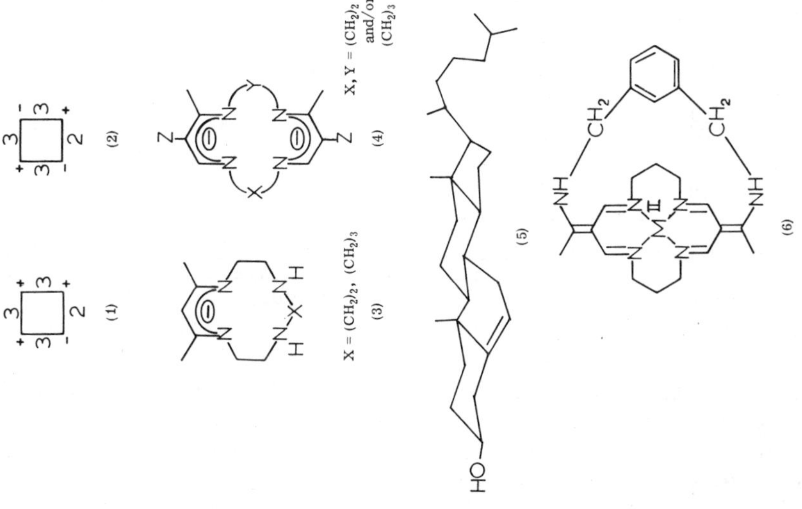

Structures

Figure 2. Saturated neutral tetraaza macro-cyclic ligands

36 ELECTROCHEMICAL STUDIES OF BIOLOGICAL SYSTEMS

Table I. Half-Wave Potentials for the Reduction of trans-Co([13-16]ane-N$_4$)Cl$_2$$^+$ in Acetonitrile Solution; Ag/AgNO$_3$ (.1M) Reference Electrode, 0.1 M n-Bu$_4$NBF$_4$ Supporting Electrolyte.

Ligand (isomer)	E$_{1/2}$ (volts)[a]	Ideal M-N (Å(	Dq^{xy} (cm^{-1})
[13]aneN$_4$	-0.66 r	1.92	2750
[14]aneN$_4$	-0.69 r	2.07	2480
[15]aneN$_4$ (I)	-0.38 r	2.28	2303
[15]aneN$_4$ (II)	-0.47 qr	2.23	2421
[16]aneN$_4$ (I)	-0.15 i	2.38	2249
[16]aneN$_4$ (II)	-0.11 i		2341

[a]Abbreviations: r, reversible; qr, quasi-reversible; i, irreversible.

derived ligand field parameter Dq^{xy} that measures the ligand field strength of the nitrogen donors of the macrocycle. This parameter is reported in the last column of Table I. On the basis that Dq^{xy} for many primary and secondary amine donors bound to Co^{3+} is ~2500 cm^{-1}, it has been concluded that [14]aneN$_4$ binds to Co^{3+} in the least strained manner (it exhibits a normal Dq^{xy} of 2480 cm^{-1}); thus, the 14-member-ed ring fits Co^{3+} best. The range of Co-N distances with the usual saturated nitrogen-containing ligands is 1.94 to 2.03 Å (8). Thus, the calculated ideal M-N distance is high by about 0.1 Å. Recognizing this limitation, ideal M-N is taken as a parameter having a best-fit value for Co^{3+}-N of 2.07.

Returning to Table I and the E$_{1/2}$ values therein, it is apparent that the value of E$_{1/2}$ is most negative for the complex in which the macro-cycle fits the metal ion best. As the misfit increases, it becomes more difficult to produce Co^{3+}; i.e., easier to reduce trans-Co(MAC)Cl$_2$$^+$. This is true even for [13]aneN$_4$, which because of its small ideal M-N might have been expected to fit Co^{3+} better than Co^{2+}. This emphasizes the fact that the relationship between ring size and redox potential is more complex than the obvious effect associated with the fact that as the oxidation state of an ion increases its size usually decreases. The data presented in Table I suggests that E$_{1/2}$ correlates with the strain energy of the Co^{3+} complex, for this does indeed increase as the misfit increases between the ideal M-N and the usual M-N distance. This is rationalized by assuming that the reduction of Co^{3+} to Co^{2+} is accom-panied by a relief in the strain energy of the macrocyclic ligand. This, in turn, is consistent with the fact that the cobalt(II) complexes of tet-raaza macrocyclic ligands are relatively labile and commonly exhibit reduced coordination numbers and distorted geometries.

On the basis of more limited data a related behavior has been reported for the Ni^{2+}/Ni^{3+} couple (Table II) (2). The 15-membered ring fits the larger, high spin Ni^{2+} ion best and this is marked by an enhanc-ed difficulty in the oxidation of Ni([15]aneN$_4$)$^{2+}$ over Ni([14]aneN$_4$)$^{2+}$ of

Table II. Half-Wave Potentials for the Oxidation of Ni(MAC)$^{2+}$ in Acetonitrile Solution, Ag/AgNO$_3$ (0.1 $\underline{M}$) Reference Electrode, 0.1 $\underline{M}$ n-Bu$_4$-
NBF$_4$ Supporting Electrolyte.

Ligand	E$_{1/2}$ (volts)[a]	Ideal M-N (Å)	Dq^{xy} (cm^{-1})	Ref.
[13]aneN$_4$	+0.7→0.9 i	1.92	---	2
[14]aneN$_4$	+0.67 r	2.07	1480	2
[15]aneN$_4$	+0.90 r	2.28	1250	2
[16]aneN$_4$	b	2.38	1110	2
Me$_2$[14]-aneN$_4$	+0.68 r	---	1471	2
Me$_4$[14]-aneN$_4$	+0.71 r	---	---	2
t-Me$_6$[14]-aneN$_4$	+0.87 r	---	1469	2
c-Me$_6$[14]-aneN$_4$	+0.86 r	---	---	9
t-Me$_6$[16]-aneN$_4$	~+1.3 i	---	---	2

[a]Abbreviations: r, reversible; qr, quasi-reversible; i, irreversible.
[b]The oxidation wave is ill-defined.

some 230 mV. Comparing ligands of similar extents of substitution,
Ni(Me$_6$[16]aneN$_4$)$^{2+}$ is much more difficult to oxidize than is Ni(Me$_6$[14]-
aneN$_4$)$^{2+}$ (a difference of about 400 mV).

A more common variety of steric strain is evident upon comparison
of the behavior of the Ni^{2+} complexes of Me$_6$[14]aneN$_4$ and [14]aneN$_4$ (2).
The presence of a gem-dimethyl group in the former, combined with the
most stable ring conformations, assures that two of the CH$_3$ groups are
oriented axially. This leads to strong repulsions between these methyls
and the monodentate solvent molecules coordinated to Ni^{3+} above and
below the plane of the macrocycle (2). Since the corresponding Ni^{2+}
complex is 4-coordinate and square planar, that interaction occurs only
for the oxidized form of the complex. The resulting destabilization of
the trivalent state causes a 200 mV increase for E$_{1/2}$ for the Ni^{2+}/Ni^{3+}
couple from +0.67 V to 0.87 V. The reliability of this conclusion is
demonstrated by the fact that the cis and trans isomers (Figure 2) of
Me$_6$[14]aneN$_4$ give closely similar values for E$_{1/2}$ (0.86 and 0.87,
respectively). Further, Me$_2$[14]aneN$_4$ and Me$_4$[14]aneN$_4$, which do not
contain gem-dimethyl groups in their 6-membered chelate rings, give
E$_{1/2}$ values very close to that of the unsubstituted ligand [14]aneN$_4$.

Complexes With Unsaturated Neutral Ligands. Figure 3 presents
the structures of a family of closely related macrocyclic ligands. All
are 14-membered rings and most have six methyl groups, three on each
of their six-membered chelate rings. The unsaturated linkages are all

associated with nitrogen donors and both the number of such linkages and their positions vary.

The Ni^{2+} complexes are all low spin and 4-coordinate, and they therefore provide a simple set of species suitable for assessing the effect of unsaturated vs saturated nitrogen donors on the relative stabilities of the various oxidation states. The change in ligand field strength upon substitution of unsaturated nitrogens into the structures is indicated by Dq^{xy}. In general, increased unsaturation enhances the ligand field strength toward high spin Ni^{2+} (such measurements are only feasible with the high spin species) (10). Only the species containing conjugated α-diimine groupings (structures 13, 15, 16, 17, and 18) add complications (2). In structures of this type, the first electron added to Ni-$(MAC)^{2+}$ goes into a π^* ligand orbital. In all other cases, one-electron reduction produces a Ni^{1+} complex and one-electron oxidation produces a Ni^{3+} species in all cases (2). It is significant that two-reduction processes occur only in the voltammograms of Ni^{2+} complexes whose ligands involve α-diimine linkages. Most generally, the advent of unsaturation in the ligands favors reduction and causes oxidation to become more difficult. In order to consider these effects more quantitatively, it is necessary to recall the results of the preceding section. $Ni(Me_6[14]4, 11\text{-dieneN}_4)^{2+}$ must be compared with $Ni(Me_6[14]aneN_4)^{2+}$ and $Ni(Me_2[14]1, 3\text{-dieneN}_4)^{2+}$ with $Ni([14]aneN_4)^{2+}$ because of the effect of axial CH_3 groups. From these comparisons it is evident that the substitution of 2 isolated imines for 2 amines in the structure increases $E_{1/2}$ for oxidation of Ni^{2+} by about 100 mV. For the conjugated case, the substitution of two imines increases $E_{1/2}$ for the same couple by about 200 mV. Thus, the α-diimine linkage stabilizes the lower state more greatly than does a pair of isolated imines. In a strictly parallel manner, $Ni(Me_6[14]1, 4, 8, 11\text{-tetraeneN}_4)^{2+}$ can be compared with $Ni(Me_6[14]aneN_4)^{2+}$ or with $Ni(Me_6[14]4, 11\text{-dieneN}_4)^{2+}$. It follows that the substitution of 4 isolated imines for amines increases $E_{1/2}$ by 180 mV and the substitution of the remaining two amines in $Me_6[14]4, 11$-dieneN$_4$ by imines increases $E_{1/2}$ by 70mV. From these considerations we offer the generalization that $E_{1/2}$ is increased by about 40-45 mV (call it 43 mV) for each amine that is replaced by an imine, so long as the imines are not conjugated. The corresponding reasoning for the conjugated case leads to the figure of +170 mV as the increase in $E_{1/2}$ for the oxidation of $Ni(MAC)^{2+}$ to the Ni^{3+} complex when a pair of amines is replaced by a conjugated α-diimine grouping. It has been pointed out that these incremental changes in $E_{1/2}$ for the Ni^{2+}/Ni^{3+} couple are roughly additive so that the $E_{1/2}$ can be predicted to a fair degree of approximation for related structures (2). A similar treatment for the reduction of $Ni(MAC)^{2+}$ is less appropriate for two reasons. The very negative reductive electrode processes are not generally so

Table III. Half-Wave Potentials[a] for the Complexes of Fe^{2+}, Co^{2+}, and Ni^{2+} with Unsaturated Tetraaza Fourteen-Membered Neutral Macrocyclic Ligands in Acetonitrile Solution, $Ag/AgNO_3$ (0.1 $\underline{M}$) Reference Electrode, 0.1 M n-Bu_4NBF_4 Supporting Electrolyte.

Ligand	No. of Isolated Imines	No. of Conjugated Imines	$E_{1/2}$(V) for Ni-$(MAC)^{2+}$ $2^+\to3^+$	$2^+\to1^+$	$2^+\to0$	Dq^{xy} (cm^{-1}) Ni^{2+}	$E_{1/2}$(V) for Fe-$(MAC)^{2+}$ $2^+\to3^+$	$2^+\to1^+$	$2^+\to0$	$2^+\to-1$	$E_{1/2}$(V) for Co-$(MAC)(CH_3CN)_2^{3+}$ $3^+\to2^+$	$3^+\to1^+$	$3^+\to0$	Dq^{xy} (cm^{-1}) Co^{3+}
$Me_6[14]4,11$-dieneN$_4$	2	0	+0.98	-1.57	---	1569	+0.44	-2.1i	---	---	+0.13	-1.69	---	2780
$Me_2[14]1,3$-dieneN$_4$	0	2	+0.86	-1.16	---	1553	---	---	---	---	+0.05	-1.17	---	2790
$Me_6[14]1,4,11$-trieneN$_4$	3	0	---	---		---	+0.51	-2.0i	---	---	---	---		
$Me_6[14]1,3,8$-trieneN$_4$	1	2	---	---		---	+0.76	-1.33	-1.85	---	---	---		
$Me_6[14]1,4,8,11$-tetraeneN$_4$	4	0	+1.05	-1.35	---	---	+0.59	-2.0i	---	---	+0.13	-1.32	---	2940
$Me_6[14]1,3,7,10$-tetraeneN$_4$	2	2	---	---		---	+0.82	-1.18	---	---	---	---		
$Me_6[14]1,3,8,10$-tetraeneN$_4$	0	4	---	---		---	+0.89	-0.80	-1.41	-1.83	---	---		
$Me_4[14]1,3,8,10$-tetraeneN$_4$	0	4	+1.00	-0.82	-1.15	1767	---	---	---	---	+0.12	-0.72	-1.61	2960
$Me_6[14]1,3,7,11$-tetraeneN$_4$	2	2	+1.05	-0.76	-1.62	---	+0.72	-1.37	---	---	---	---		

[a] Abbreviations: i, irreversible.

well behaved and the added electron is centered on the metal atom in some cases and on the ligand in others.

In most general terms, the redox behavior of the iron (II) complexes (11) parallels that of the nickel derivatives that have just been considered. Increasing unsaturation favors the lower state and the presence of α-diimine linkages is accompanied by multiple reduction processes. There are, however, significant differences. The Fe^{2+}/Fe^{3+} couple is substantially more sensitive to unsaturation than the Ni^{2+}/Ni^{3+} couple (Table III) and the one-electron reduction product of $Fe(MAC)^{2+}$ is best formulated as an Fe^{1+} derivative even in the complexes of ligands containing α-diimine linkages (12). This has been thoroughly demonstrated by the isolation and characterization of the salts of the complex $Fe(Me_6[14]1,3,8,10\text{-tetraeneN}_4)^{1+}$. It should be emphasized that all of the iron(II) complexes are solvated in acetonitrile solutions (13). In order to facilitate comparisons, $E_{1/2}$ values (14) for Fe^{2+} complexes with several saturated tetraaza macrocycles are summarized in Table IV.

Table IV. Half-Wave Potentials for the Complexes of Fe^{2+} with Saturated Tetraaza Macrocyclic Ligands in Acetonitrile Solution, Ag/AgNO$_3$ (0.1 $\underline{M}$) Reference Electrode, 0.1 $\underline{M}$ n-Bu$_4$NBF$_4$, Supporting Electrolyte.

Ligand	$E_{1/2}$ (volts)[a]	
	+2 → +3	+2 → +1
[14]aneN$_4$	+.24	----
[15]aneN$_4$	+.49 qr	-2.3 i
[16]aneN$_4$	+.66 i	-2.1 i
Me$_2$[14]aneN$_4$	+.27	-2.2 i
Me$_6$[14]aneN$_4$	+.38	-2.1 i

[a] Abbreviations: qr, quasi-reversible; i, irreversible.

From Table IV the same effects are evident for iron complexes as for nickel complexes; i.e., increasing ring size favors the lower state as does the presence of axial CH$_3$ groups. Now, comparing these values (Table IV) with those in Table III for unsaturated ligands, the following results are found (averaged to fit the several cases reasonably well): $\Delta E_{1/2}$ = +140 mV for a pair of axial CH$_3$ groups, +300 mV for an α-diimine group, and +50 mV for one isolated C=N. From these values, it is apparent that the enhanced sensitivity of the Fe^{2+}/Fe^{3+} couple to the presence of ligand unsaturation resides in the rather large effect of the α-diimine group. This can be attributed to the special stability of the Fe^{II}-α-diimine ring structure (15).

In view of the systematic structure-potential relationships observed for the Ni^{2+}/Ni^{3+} and Fe^{2+}/Fe^{3+} couples, it is surprising that the Co^{2+}/Co^{3+} couple is insensitive to variations in the degree of ligand unsaturation (16). A similar observation (17) has been made on the complexes $Co(MAC)(H_2O)_2^{3+}$. This is perhaps the result of the fact that the orbital receiving the added electron (d_{z^2}) is oriented on the z axis and relatively unaffected by the in-plane ligand field. Further, since the unsaturated macrocyclic ligands are relatively inflexible, as compared to the saturated tetraaza macrocyclic ligands discussed earlier, little rearrangement is likely to occur upon reduction. The studies by Endicott et al (18) on the electrode potentials of the Co^{3+}/Co^{2+} couple in complexes containing a constant macrocyclic ligand and a variety of axial ligands supports this view. A strong dependence of the potential on the nature of the monodentate axial ligands was observed. These variations are related to the ligand field strengths of the axial ligands, a result that is easily rationalized if the receptor orbital on Co^{3+} is antibonding σ^* (d_{z^2}).

The Co^{2+}/Co^{1+} couple is responsive to the extent and nature of the unsaturation of the ligands listed in Table III. $E_{1/2}$ increases from -2.3 V for [14]aneN$_4$ (7) to -1.69 V for Me$_6$[14]4,11-dieneN$_4$ and -1.32 V for Me$_6$[14]1,4,8,11-tetraeneN$_4$ (Table III). The effect of α-diimines is more extreme with $E_{1/2}$ for the Me$_2$[14]1,3-dieneN$_4$ derivative being -1.17 V and that for Me$_4$[14]1,3,8,10-tetraeneN$_4$ being -0.72 V. Since the very negative $E_{1/2}$ for the saturated ligand system bears a large uncertainty, the magnitudes of these shifts are assessed by comparing values for the unsaturated ligand derivatives. On that basis, $\Delta E_{1/2} \approx$ +190 mV for each isolated C=N group and ~450 mV for each α-diimine grouping. The greater sensitivity for the Co^{2+}/Co^{1+} couple as compared to the isoelectronic Ni^{3+}/Ni^{2+} couple is attributed to the lower valence states involved in the former case. If back-bonding is a consideration it should be more important for lower states.

Complexes With Anionic Tetraaza Macrocyclic Ligands. Ligands of general structures 3 and 4 have been utilized to characterize the oxidation-reduction behavior of complexes with macrocyclic anionic ligands. The Ni^{2+} complexes of ligands of structure 3 exhibit irreversible oxidations at modest potentials ($E_{1/2} \cong$ +.27 V for X = (CH$_2$)$_2$ and +.23 V for X = (CH$_2$)$_3$ and very cathodic reductions (-2.30 V and -2.34 V, respectively (2)). These data show that the presence of charge on the ligand facilitates the oxidation process since the least positive values observed for this couple with neutral macrocyclic ligands is some 400 mV more positive. Also, reductions are made extremely unfavorable and the effect of ring size on $E_{1/2}$ has become very small. The complexes of structure 4, having Z = H and ring sizes of 14, 15, and 16 members, show irreversible oxidations at $E_{1/2}$ values of approximately

-0.34 V, -0.39 V and -0.29 V, respectively. Not only are these values extremely cathodic for the Ni^{2+}/Ni^{3+} couple, but greater uncertainties arise in their values because of their irreversibility. These uncertainties and the proximity of their approximate $E_{1/2}$ values, suggest further that the oxidation reaction is not sensitive to ring size among these structures. This insensitivity to ring size is supported by data for Ni^{2+} complexes of structure 4 having Z = -C⟨$^O_{CH_3}$ (Table V (2)).

The data of Table V has been obtained under the same conditions as those for all the systems described above (19). As will be apparent, shortly, the derivatives of the ligands having structure 4 are more readily studied in a different solvent; however, the limited data available from measurements in acetonitrile solutions is quite revealing. Early

Table V. Half-Wave Potentials for Ni^{2+} Complexes of Dianionic Ligands of Structure 4, Having Different Substituents (Z) in Acetonitrile Solutions, Ag/AgNO$_3$ (0.1 M) Reference Electrode, 0.1 M n-Bu$_4$NBF$_4$ Supporting Electrolyte.

| Ring Size | Substituent | $E_{1/2}$ (volts)[a] | | |
		NiL→NiL$^+$	NiL→NiL^{2+}	NiL → NiL$^-$
15	H	-0.39 i	+1.77 i	-2.77 i
15	-C⟨$^O_{CH_3}$	+0.26	+0.9 i	-2.53 i
14	-C⟨$^O_{CH_3}$	+0.25	+0.97 i	---
15	-C⟨^{O}⟨O⟩CH$_3$	+0.24	+0.85 i	-2.38 qr
15	-C⟨^{O}⟨O⟩	+0.27	+0.89 i	-2.39 qr
15	-C⟨^{O}⟨O⟩NO$_2$	+0.37	+0.96	-1.42 -1.98 qr
15	-NO$_2$	+0.64	+1.08	-2.01 qr -2.5 i

[a]Abbreviations: i, irreversible; qr, quasi-reversible.

studies (2) were carried out on the 14- and 15-membered ring species of structure 4 having acetyl groups as their substituents, Z. Product studies (2) showed that the first oxidation reaction yields a square planar Ni^{3+} complex. This process is electrochemically reversible for all the complexes of Table V except the parent compound (Z = H). In that case, it is probable that oxidation at the metal center is quickly followed by one or more ligand reactions. It has been shown that complexes having chelate rings of this class undergo a coupling reaction upon oxidation (20,21).

Though the absolute value for $E_{1/2}$ suffers from considerable uncertainty, it remains clear that the first oxidation reaction of the complex with the unsubstituted ligand occurs at a highly cathodic potential. Further, the replacement of the hydrogen atom by electron withdrawing substituents shifts the potential for this process dramatically in the anodic direction. In fact, the range of $E_{1/2}$ values reported for different derivatives in Table V spans approximately one volt. Thus, the effect of substituents on the metal-centered oxidation is truly remarkable. The magnitude of incremental change in $E_{1/2}$ with substituent correlates linearly with the Hammet (22) substituent constants (σ_p) for the three benzoyl derivatives (the slope (ρ) = 0.138). This emphasizes the fact that the substituent effect is felt directly by the metal ion.

It has been possible to demonstrate the substituent effect more convincingly using data obtained on dmf solutions (Table VI,(19)). As

Table VI. Half-Wave Potentials for Ni^{2+} Complexes of Dianionic Ligands of Structure 4, Having Different Substituents (Z), in Dimethylformamide Solutions, $Ag/AgNO_3$ (0.1 M) Reference Electrode, 0.1 M n-Bu$_4$NBF$_4$ Supporting Electrolyte.

Compound Number	Substituent	Oxidation[a] $E_{1/2}$ (volts)	Reduction[a] $E_{1/2}$ (volts)	d-d Band γ(kK) (ϵ)
1	$-CH_2CH_2C\overset{O}{\underset{OC_2H_5}{}}$	(1)-0.39 (2)+0.29 i	-2.92 qr	17.02 (224)
2	-H	(1)-0.36 i (2)+0.39 i	-2.83 qr	17.15 (169)
3	$CH_3C\overset{O}{<}$; -H	(1) -0.14 i (2)+0.51 i	-2.64 qr	18.32 (197)
4	$-C\overset{O}{\underset{NH-\alpha-C_{10}H_7}{}}$	(1)+0.03 (2)+0.55 i	-2.59 qr	18.66 (285)
5	$-\overset{O}{\underset{C_6H_5}{C}}$	(1)+0.21 (2)+0.64 i	-2.42 qr	19.27 (433)
6	$\overset{O}{\underset{CH_3}{C}}$	(1)+0.22 (2)+0.64 i	-2.52 qr	19.30 (330)
7	$-NO_2$	(1)+.42 (2)+1.04 i	(1)-2.00 (2)-2.60 i	19.80 (922)

[a] Abbreviations: qr, quasireversible; i, irreversible.

stated earlier, the first oxidation wave has been assigned to the $Ni^{2+}/-Ni^{3+}$ couple. It is generally well behaved except, again, for those compounds having hydrogen atoms on the γ-carbons. The second oxidation has been attributed to ligand oxidation (2), a reaction that has been well demonstrated using chemical oxidizing agents (23, 24). These data indicate that electron withdrawing substituents are highly effective at producing more positive potentials while alkyl substituents produce potentials at least as negative as that of the parent compound (Z = H). The total range of potentials spanned by these substituted derivatives in dmf is 0.81 V. The responsiveness of $E_{1/2}$ to substitution is shown most clearly by the linear free energy relationship of Figure 4. $E_{1/2}$ correlates well with σ_p, σ_p^-, and σ_m and the quality of the correlation is displayed in Table VII. The correlations with σ_p and σ_p^- are equally good and

Table VII. Linear Free Energy Correlation[a] for $E_{1/2}$.

σ	$\rho_{E_{1/2}}$	$(E_{1/2})_{\sigma=0}$	r
σ_p^-	.645	-.362	.994
σ_p	1.013	-.345	.993
σ_m	1.016	-.331	.978

[a]The appropriate equation is

$$E_{1/2} = \rho_{E_{1/2}} \sigma + (E_{1/2})_{\sigma=0}$$

that with σ_m is not much weaker. This suggests that resonance interactions are not principally responsible for the correlation.

These data establish the fact that the effects of substituents on ligands of structure 4 are transmitted strongly to the metal ion and that the redox properties of the metal ion can be controlled with some precision by the choice of substituents. These relationships are supported further by the parallel correlation of the energies (cm^{-1}) of the low-energy electronic transitions of these complexes (Table VI) with the same sets of substituent constants (Table VIII, Figure 5). These spectral bands are presumed to involve d-d transitions because they fall in the usual range and exhibit extinction coefficients that are typical of such electronic processes for low spin, square planar Ni^{II} complexes (25). The quality of the correlations follows that for $E_{1/2}$ very closely: $\sigma_p^- \approx \sigma_p > \sigma_m$. Assuming the spectral bands to be of d-d origin, ligand field strength varies with band position. It follows that the substituent effect is exerted by alterations in ligand field strength, a relationship that is certainly not unexpected.

As one looks toward possible applications of these results in the design of biochemical models, $E_{1/2}$ is a functional property; i.e., the ability of the metal ion to function in certain model systems will depend on achieving appropriate $E_{1/2}$ values. On the other hand, electronic

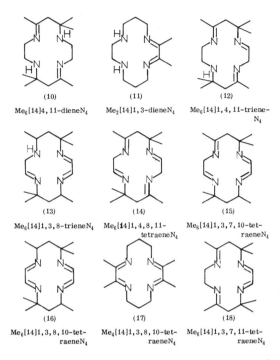

(10)

Me₆[14]4,11-dieneN₄

(11)

Me₂[14]1,3-dieneN₄

(12)

Me₆[14]1,4,11-triene-
N₄

(13)

Me₆[14]1,3,8-trieneN₄

(14)

Me₆[14]1,4,8,11-
tetraeneN₄

(15)

Me₆[14]1,3,7,10-tet-
raeneN₄

(16)

Me₆[14]1,3,8,10-tet-
raeneN₄

(17)

Me₄[14]1,3,8,10-tet-
raeneN₄

(18)

Me₆[14]1,3,7,11-tet-
raeneN₄

*Figure 3.   Unsaturated tetraaza fourteen-membered
neutral macrocyclic ligands*

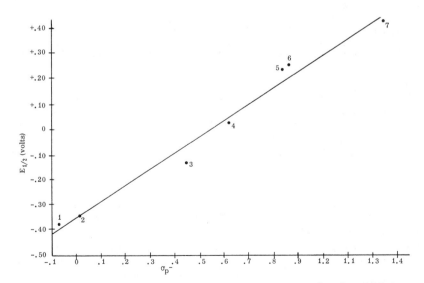

Figure 4.   Correlation of $E_{1/2}$ with σ_p^- for nickel complexes listed in Table 6

transition energies and ligand field strengths are diagnostic of the electronic structure of the metal ion. It is significant that the functional property and the diagnostic property both respond linearly to substituent effects.

Table VIII. Linear Free Energy Correlation[a] for the Low Energy Electronic Spectral Bonds.

σ	ρ_{d-d}	$(E_{d-d})_{\sigma=0}$	r
σ_p^-	2.205	17.243	.991
σ_p	3.508	17.308	.987
σ_m	3.947	17.371	.958

[a]The appropriate equation is

$$E_{d-d} = \rho_{d-d}\sigma + (E_{d-d})_{\sigma=0}$$

The ligands to which the most attention has been devoted here are of special interest in on-going programs directed toward the developing of relatively uncomplicated molecules and complex ions that can mimic the function of heme proteins (19, 23, 26). Progress in that direction has included the syntheses of the iron complexes of ligands of structure 4 as the prosthetic groups of such model systems (27,28). The electrochemical behavior of these species is shown in Table IX. The complexes with the unsubstituted ligands are extremely sensitive to oxidation and several ligand oxidation waves are observed at $E_{1/2}$ values more positive than the highly coulombic Fe^{2+}/Fe^{3+} couple. The oxidation wave associated with the Fe^{2+}/Fe^{3+} couple was identified by esr

Table IX. Half-Wave Potentials for the Iron(II) Complexes of Ligands of Structure 4, in Acetonitrile Solution, $Ag/AgNO_3$ (0.1 $\underline{M}$) Reference Electrode, 0.1 $\underline{M}$ n-Bu_4NBF_4 Supporting Electrolyte.

Complex	$E_{1/2}$, Oxidations (volts)				$E_{1/2}$, Reductions (volts)
	Fe^{2+}/Fe^{3+}	Ligand			
$Fe(Me_2[14]tetraenatoN_4)$	-0.77	+0.02 i	~0.7	~1.2	-2.38
$Fe(Me_2[15]tetraenatoN_4)$	-0.89	+0.03 i	~0.7	~1.2	-2.47 qr
$Fe(Me_6[15]tetraenatoN_4)$	-0.90	+0.30 i	~0.8		-2.57
$Fe(Me_2[16]tetraenatoN_4)$	-0.83 i	~1.2 i			

measurements. The extremely cathodic values for this couple (-0.7 → -0.9 V) are consistent with the data reported above for the Ni^{2+}/Ni^{3+} couple. The ligand oxidations occurring at low positive potentials give well-defined but irreversible waves while the additional oxidation waves are poorly defined. The 16-membered ring derivative fails to exhibit the low energy ligand oxidation. The electrochemical behavior of this species is also unusual in that the Fe^{2+}/Fe^{3+} couple is irreversible and esr studies on the Fe^{2+}/Fe^{3+} oxidation product failed to reveal a signal for Fe^{3+}. It is suspected that the first ligand oxidation is coupled to the metal oxidation in this case. Such processes are well documented from synthetic studies on iron complexes (29). The complex denoted in Table IX as $Fe(Me_6[15]tetraenatoN_4)$ has its four additional methyl groups as substituents on the carbon atoms of the five-membered chelate rings.

Studies on iron complexes of substituted ligands of structure 4 remain in the synthetic stages at this point; however, the opportunities that these investigations should provide can be anticipated by a number of results that are presently available. In order for an iron atom to function in a heme protein, two general structural assemblages must meet critical requirements. The first is the coordination sphere of the metal ion, which must cause the functional properties (e.g., $E_{1/2}$) to fall in useable ranges. The second involves the noncoordinated structural array in the vicinity of the metal site--we refer to this as the associated proximate structure (APS). The latter may serve one or more of several functions, such as providing propitiously oriented functional groups or determining the polarity in that region. For example, to facilitate O_2 transport by Fe^{2+}, it is generally presumed that the APS must both provide a nonpolar, aprotic environment and prevent the close approach of two iron atoms. In initial experiments designed to provide a suitable APS while using the substituent-based control of $E_{1/2}$ described immediately above, we have formed a cholesterol derivative (30) of the nickel complex $Ni(Me_2[15]tetraenatoN_4)$ by first forming a derivative with the substituents Z (structure 4) as succinoyl half-esters $-\overset{O}{\overset{\|}{C}}CH_2CH_2\overset{O}{\overset{\|}{C}}OCH_3$ and then carrying out a trans-esterification reaction with cholesterol (structure 5). It would be expected that this large hydrophobic moiety would orient itself over the metal ion in aqueous or other highly polar media, forming a nonpolar umbrella (the APS) in the vicinity of the metal ion. The nickel(II) complex having this bulky substituent has been thoroughly characterized (29) and it exhibits electrochemical properties closely similar to those of the complex having simple acetyl substituents. Oxidations: $E_{1/2}$ = +0.21 V; +.67 V(i); reduction: -2.48 V(qr)--compare to Table VI. This clearly establishes the opportunity to design structures having both the requisite coordination spheres and associated proximate structures to mimic heme proteins with totally synthetic systems.

A closer approach to the totally synthetic analogue of an heme protein--O_2 binding site, under study in these laboratories (31), makes use of a more complicated modification of structure 4. The final product contains a ligand having an organic moiety sheltering the metal ion site but fixed in position by two points of attachment (structure 6).

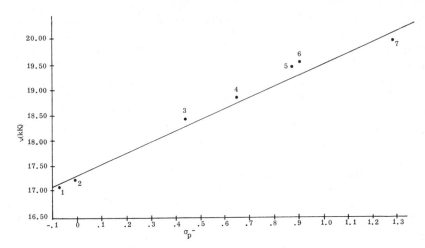

Figure 5. Correlation of d-d transition energies (kK) with σ_p^- for nickel complexes in Table 6

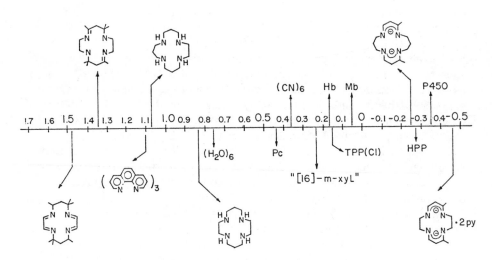

Figure 6. Half-wave potentials for the Fe^{2+}/Fe^{3+} couple in selected complexes adjusted to SHE reference electrode

Structure 6 as drawn is misleading in that it does not properly represent the orientation of the m-xylyl group. This group is constrained to occupy a position above the metal ion site producing a hydrophobic cavity (the "dry cave") in that location. The coordination site on the opposite side of the coordination plane is available for binding of monodentate ligands and the N-methylimidazole adduct has been characterized (31). The Fe^{2+}/Fe^{3+} couple for this species displays an $E_{1/2}$ of -0.36 V (vs Ag/AgNO$_3$ (0.1 M) in acetonitrile (usual conditions). Figure 6 shows how this process for the new complex, identified as "[16]-m-xyl", relates to those for a number of heme proteins, the common iron complexes and a few other complexes with macrocyclic ligands. It is particularly encouraging that the $E_{1/2}$ value for this first totally synthetic dry cave complex of iron(II) falls in close proximity to that of hemoglobin. For this comparison, the values measured in acetonitrile vs a Ag/AgNO$_3$ (0.1 M) reference electrode have been recalculated to the standard aqueous hydrogen electrode by the addition of 0.58 V (1). Preliminary evidence (30) indicates that the dry cave complex does interact with the small molecules CO and O$_2$.

LITERATURE CITED

1. To adjust $E_{1/2}$ from SCE to Ag/AgNO$_3$ (0.1 M) reference electrode in acetonitrile, 0.334 V is subtracted (Mann, C., and Barnes, K., "Electrochemical Reactions in Nonaqueous Systems," Marcel Dekker, New York, 1970). Normal hydrogen electrode is not stable in acetonitrile but appropriate adjustment from Ag/AgNO$_3$ (0.1 M) to SHE appears to involve addition of 0.58 to 0.63 V (Butler, J. N., Advan. Electrochemical Engr., (1970), 7, 87). To adjust from SCE in dmf to Ag/AgNO$_3$ (0.1 M) in acetonitrile, subtract 0.354 V (Butler, p. 135 and above). To adjust from SCE in dmso to Ag/AgNO$_3$ (0.1 M) in acetonitrile, add 0.016 volt (Mann and Barnes). To adjust from SCE in benzonitrile to Ag/AgNO$_3$ (0.1 M) subtract 0.454 volt (Mann and Barnes, p. 479).

2. Lovecchio, F. V., Gore, E. S., and Busch, D. H., J. Am. Chem. Soc., (1972), 96, 3109.

3. Hung, Y., Ph.D. Thesis, The Ohio State University, 1976.

4. DeHayes, L. J., and Busch, D. H., Inorg. Chem., (1973), 12, 1505.

5. DeHayes, L. J., and Busch, D. H., Inorg. Chem., (1973), 12, 2010.

6. Martin, L. Y., DeHayes, L. J., Zompa, L. J., and Busch, D.H., J. Am. Chem. Soc., (1974), 96, 4046.

7. Hung, Y., Jackels, S., and Busch, D.H., submitted for publication.

8. Saito, Y., "Spectroscopy and Structure of Metal Chelate Compounds," Ed. by K. Nakamoto and P. J. McCarthy, John Wiley, New York, 1968.

9. Olson, D.C., and Vasilevskis, J., Inorg. Chem., (1969), 8, 1611.

10. Busch, D.H., Helv. Chim. Acta, (1967), Werner Commemoration Volume, 174.

11. Dabrowiak, J.C., Lovecchio, F. V., Goedken, V. L., and Busch, D. H., J. Am. Chem. Soc., (1972), 94, 5502.

12. Rakowski, M.C., and Busch, D.H., J. Am. Chem. Soc., (1975), 97, 2570.

13. Watkins, D. D., Riley, D.P., Stone, J.A., and Busch, D. H., Inorg. Chem., (1976), 15, 387.

14. Rakowski, M.C., Ph.D. Thesis, The Ohio State University, 1974.

15. Figgins, P. E., and Busch, D. H., J. Phys. Chem., (1961), 65, 2236.

16. Tait, A.M., Lovecchio, F.V., and Busch, D. H., submitted for publication.

17. Rillema, D. P., Endicott, J. F., and Patel, R. C., J. Am. Chem. Soc., (1972), 94, 394.

18. Rillema, D. P., Endicott, J. F., and Popaconstantineu, E., Inorg. Chem., (1971), 10, 1739.

19. Pillsbury, D. G., and Busch, D. H., submitted for publication.

20. Cunningham, J.A., and Sievers, R.E., J. Am. Chem. Soc., (1973), 95, 7183.

21. Dabrowiak, J.C., private communication.

22. Gordon, A.J., and Ford, R.A., "The Chemist's Companion," John Wiley, New York, pp. 144-153, 1972.

23. Truex, T., and Holm, R. H., J. Am. Chem. Soc., (1972), 94, 4529.

24. Hipp, C. J., Lindoy, L. F., and Busch, D.H., Inorg. Chem., (1972), 11, 1988.

25. Lever, A.B.P., "Inorganic Electronic Spectroscopy," Elsevier Publishing Co., Amsterdam, 1968.

26. Koch, S., Tang, S.E., and Holm, R.H., J. Am. Chem. Soc., (1975), 97, 914.

27. Riley, D.P., Stone, J.A., and Busch, D.H., J. Am. Chem. Soc., in press.

28. Koch, S., Holm, R.H., and Frankel, R., J. Am. Chem. Soc., (1975), 97, 6714.

29. Dabrowiak, J.C., and Busch, D.H., Inorg. Chem., (1975), 14, 1881 and references therein.

30. Pillsbury, D.G., and Busch, D.H., unpublished results.

31. Schammel, W. P., Ph.D. Thesis, The Ohio State University, 1976.

Electrochemical Studies on the Thermodynamics of Electron Transfer and Ligand Binding of Several Metalloporphyrins in Aprotic Solvents

K. M. KADISH,* L. K. THOMPSON, D. BEROIZ, and L. A. BOTTOMLEY

Department of Chemistry, California State University, Fullerton, Calif. 92634

The utility of electrochemistry as a tool for studying physical properties of metalloporphyrins has been recognized extensively during the last several years ($\underline{1},\underline{2}$). It is now well known that the half wave potentials for metalloporphyrin electro-oxidation-reduction are directly influenced by the number and type of complexed axial ligands and that these may, in some instances, be related to the dioxygen carrying ability of the M(II) metalloporphyrin where M is Co($\underline{3}$-$\underline{6}$), Fe ($\underline{7}$-$\underline{9}$) or Mn ($\underline{10}$).

Enthalpy and entropy values for complexation of cobalt(II) ($\underline{4},\underline{11},\underline{12}$) and iron(II) ($\underline{13}$) porphyrins, by several Lewis bases have been reported, but similar data for Lewis base complexation is not available for the oxidized cobalt(III) and iron(III) species. This data is of some interest in that changes of solvation and/or ligand binding, concomitant with electron transfer, may produce large entropic effects which would significantly shift the half wave potentials as a function of temperature and could alter reported relationships between half wave potentials at 20°C and stability constants of dioxygen complexes elucidated at reduced temperatures.

Therefore, we have undertaken in our laboratory a systematic study of the thermodynamics associated with electron transfer to and from metalloporphyrins in nonaqueous media. In this paper we present initial results on the entropy changes associated with π cation and π anion radical reactions of several porphyrin complexes. We have compared these to entropy changes for electron transfer to and from the central metal of Co(II)TPP and Fe(II)TPP in bonding and nonbonding solvents. Thermodynamic data is also presented for Lewis base addition to Co(II)TPP and Fe(II)TPP to form the mono and bis pyridine complexes, respectively.

The method of investigation consisted of measuring each reversible half wave potential, $E_{\frac{1}{2}}$, at several temperatures, and from a plot of $E_{\frac{1}{2}}$ vs T, the entropy was approximated utilizing the Gibbs-Helmholtz equation: ($\underline{14}$)

*Present address: Department of Chemistry, University of Houston, Houston, Texas 77004

$$\Delta S \simeq nF \frac{\Delta E_{\frac{1}{2}}}{(\Delta T)_p} = \frac{\Delta H - \Delta G}{T} \qquad (1)$$

where ΔS, ΔH and ΔG are the entropy, enthalpy and free energy of the electron transfer step and $E_{\frac{1}{2}}$ is the polarographic half wave potential. The validity of this equation involves the assumption that $E° \simeq E_{\frac{1}{2}}$ as measured by cyclic voltammetry, that the diffusion coefficients of the oxidized and reduced forms of the complex are about equal and that the ambient pressure is constant. For the compounds and conditions of this study, all of these assumptions are valid.

The half wave potentials are also a function of axial ligand complexation of both the oxidized and reduced species and may be described by equation 2 for electroreductions. (15)

$$(E_{\frac{1}{2}})_c = (E_{\frac{1}{2}})_s - \frac{RT}{nF} \ln \frac{K_{ox}}{K_{red}} - \frac{RT}{nF} \ln (L)^{p-q} \qquad (2)$$

$(E_{\frac{1}{2}})_c$ and $(E_{\frac{1}{2}})_s$ are the half wave potentials of the complexed and uncomplexed oxidized species, respectively; K_{ox} and K_{red} are the formation constants of the oxidized and reduced complex, (L) is the free concentration of the complexing ligand, p and q are the number of ligands bound to the oxidized and reduced species, and n is the number of electrons transferred in the diffusion controlled reaction.

It is seen from equation 2 that the more stable the oxidized complex, i.e. the larger the K_{ox} relative to K_{red}, the more cathodic will be its reduction potential. The half wave potential should also shift with changing concentration of the ligand by $-(p-q)RT/nF \ln(L)$. This relationship allows us to determine the coordination number of the complex after we have calculated the number of electrons transferred. Insertion of p and q into equation 2 will yield the formation constant, which, when determined at several temperatures, then allows us to calculate ΔH and ΔS for ligand binding using the Van't Hoff equation. (14)

Experimental

Chemicals. The porphyrins CoTPP, FeTPPCl, H_2TPP, ZnTPP, MgTPP, and MnTPPCl were purchased from Strem Chemical Inc. (Danvers, Mass.) and were used as received. Reagent grade N,N-dimethylformamide (DMF) and butyronitrile were dried over activated 4Å molecular sieves before use. Other solvents were spectral or reagent grade and were used as received. The supporting electrolyte, tetrabutylammonium perchlorate (TBAP), was recrystallized twice from absolute methanol and dried at reduced pressure over P_4O_{10}.

Instrumentation and Data Analysis. Cyclic voltammograms
were obtained on a PAR 174 polarographic analyzer in conjunction
with an X-Y recorder. Measurements were made in a Brinkman Model
EA 875-5 electrochemical cell which was immersed in a 400 ml Dewar
or a 1000 ml beaker and allowed to equilibrate before commensing
the experiments. Temperatures below 0°C were achieved through the
use of slush baths containing KCl and ice. Temperatures above
0°C were maintained by immersion of the cell into a temperature
bath which was accurate to 0.5° during the time of the measurement.

A three electrode system consisting of two platinum electrodes and a standard calomel electrode (SCE) was used. The SCE
was separated from the bulk of the solution by a double bridge containing fine glass frits and filled with solvent and supporting
electrolyte. Solutions in the bridge were changed periodically
to avoid aqueous contamination from the SCE. The reference
electrode was placed into the upper bridge after temperature
equilibration in the electrochemical cell and removed after each
scan in order to re-equilibrate at room temperature for 5 minutes
between runs. The potential was periodically checked against a
second SCE and was found not to vary by more than 1mV between
temperature trials.

Solutions were purged of oxygen by degassing with purified
nitrogen before running cyclic voltammograms. After deaeration,
a blanket of nitrogen was kept over the solution. The half wave
potentials were measured as the average of the anodic and the
cathodic peak potentials and ΔS was calculated by means of a
least squares best fit program which analyzed $dE_{1/2}/dT$. Calculations of ΔS are based on the average of 3 - 5 measurements. ΔH,
ΔS, and ΔG from the Van't Hoff plots were also analyzed by a
least squares best fit program. Uncertainties are expressed as
the relative average deviation of the measurements.

RESULTS AND DISCUSSION

Entropy of Electron Transfer Reactions at Porphyrin Ring. A
typical cyclic voltammogram at 295°K is shown in Figure 1 for
ZnTPP in butyronitrile. The four electrode reactions are all
reversible and have been shown to correspond to formation of a π
anion radical and dianion product at -1.32 V and -1.71 V(16) and
a π cation radical and dication at +0.81 V and +1.10 V.(17). The
potentials for each reaction were not constant but shifted along
the potential axis as a linear function of temperature. A plot of
$E_{1/2}$ vs T was constructed (Figure 2) and using equation 1, ΔS for
the initial one electron addition and one electron abstraction
from the neutral ZnTPP was calculated as 9.3±1.5 eu and 4.0±0.3
eu, respectively. Entropy changes for other π radical reactions
were calculated and are summarized in Table I.

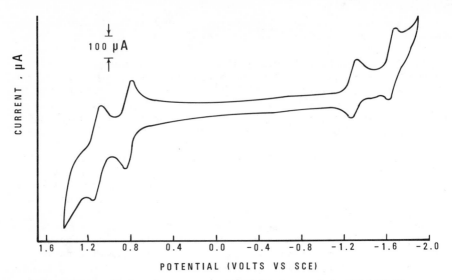

Figure 1. Cyclic voltammogram of ZnTPP in butyronitrile, 0.1M TBAP

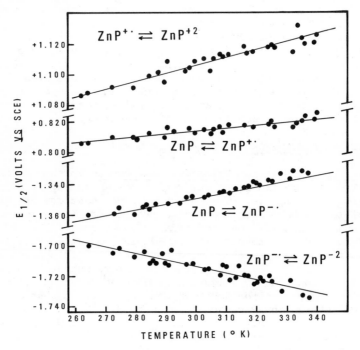

*Figure 2. Half wave potential as a function of temperature for the
four electrode reactions of ZnTPP in butyronitrile, 0.1M TBAP*

Table I

Entropy data for oxidation or reduction of a neutral porphyrin ring to yield π anion and π cation radicals in butyronitrile, 0.1M TBAP

Porphyrin	Ring Oxidation (a)		Ring Reduction(b)	
	$E_{\frac{1}{2}}$ (V)	ΔS(eu)	$E_{\frac{1}{2}}$ (V)	ΔS(eu)
H_2TPP	1.09	5.0 ± 0.6	-1.14	-3.3 ± 0.2
Zn(II)TPP	0.81	4.0 ± 0.3	-1.35	9.3 ± 1.5
Mg(II)TPP	0.69	13.0 ± 1.2	-1.56	-4.0 ± 0.6
Mn(II)TPP	(c)	(c)	-1.50	-6.8 ± 2.0

(a) Corresponds to MTPP $\rightleftarrows$ MTPP$^{+\cdot}$ + e$^-$
(b) Corresponds to MTPP + e$^-$ $\rightleftarrows$ MTPP$^{-\cdot}$
(c) Oxidation of central metal occurs before ring oxidation

Provided that the initial reactant is neutral and the entropy change due to the electron transfer step is larger than that due to solvent reorganization and change in ligand coordination, it will be observed that ΔS is positive for oxidations and negative for reductions(18). Although the data in Table I is limited to only a few compounds, the following observations may be made: porphyrin ring oxidations to yield π cation radicals involve a positive entropy change, while porphyrin ring reductions to yield π anion radicals, with the exception of ZnTPP, involve a negative entropy change.

Similar positive values of ΔS were obtained for reduction of ZnTPP in DMF and DMSO. The ΔS = 9.3±1.5 eu in butyronitrile represents an increase of 13 to 16 eu over the other neutral porphyrin reductions in Table I, and might be accounted for by loss of an axial ligand and a change from 5 to 4 coordinate geometry upon reduction. An alternate explantion of the large positive ΔS might be that ZnTPP exists as a loosely held dimer in butyronitrile which splits on reduction to yield the π anion radical. However, this explanation does not seem likely since no evidence for dimerization has been observed for neutral ZnTPP, which has been extensively studied by spectroscopic methods (19,20) and electrochemical methodologies(16,17). In contrast, however, zinc etioporphyrin and zinc octaethylporphyrin readily form dimers at low temperature in non-bonding solvents and in

methylcyclohexane a ΔS of -17.8 and -21.5 eu has been calculated
for dimerization of these complexes. (21)

Electrode Reactions of CoTPP. Entropy changes are summarized in Table II for the one electron reduction of CoTPP to yield a
negatively charged Co(I) complex.

$$Co(II)TPP + e \rightleftharpoons [Co(I)TPP]^- \tag{3}$$

as well as for the single electron oxidation to yield a positively charged Co(III) complex.

$$Co(II)TPP \rightleftharpoons [Co(III)TPP]^+ + e \tag{4}$$

In the absence of nitrogenous bases, the half wave potential for reaction 3 is only slightly dependent on solution conditions. Half wave potentials at room temperature vary by less
than 80 mV on going from a solvent of low complexing ability such
as butyronitrile or CH_2Cl_2 to a solvent of higher complexing
strength such as DMF or DMSO. (22) Additions of butyronitrile to
toluene solutions of Co(II)TPP do not result in spectral changes
indicative of complexation (23), while titration of DMSO into
butyronitrile solutions of Co(II)TPP does not change the half
wave potential for Co(II) reduction. Both of these experiments
indicate that a change in solvent coordination does not occur at
the axial position of cobalt(II).

Only in the presence of strong ligands will cobalt(II)TPP
react to form mono and bis ligand adducts. These interpretations
are confirmed by the data in Table II. Reduction of Co(II)TPP in
either butyronitrile or DMSO gives a zero or small positive ΔS
for reaction 3, while in DMSO, which was 1.13 $\underline{M}$ in pyridine $\Delta S =$
16.2 ± 2.4 eu. Since Co(I)TPP does not bind axial ligands (24), a
positive ΔS is indicative of ligand release upon reduction of
Co(II).

Pyridine titrations of Co(II)TPP - DMSO solutions, when
followed electrochemically, gave a $\Delta E_{\frac{1}{2}}/\Delta \log(\text{pyridine}) = -60mV$
between 0.5 and 3 M pyridine. (25) This slope is a clear indica-
tion that one more pyridine is complexed by Co(II) than by
Co(I), so that the electrode reaction in DMSO, 1.13 $\underline{M}$ pyridine
may be formulated as:

$$[Co(II)TPP\cdot Py] + e \rightleftharpoons [Co(I)TPP]^- + Py \tag{5}$$

for which the overall $\Delta S = 16.2 \pm 2.4$ eu. This includes both the
electron transfer step and the coupled ligand dissociation.

Combining the inverse of reaction 5 with reaction 3 yields
by a Hess's law relationship the overall reaction for ligand
addition given by equation 6,

$$Co(II)TPP + Py \rightleftharpoons Co(II)TPP\cdot Py; \tag{6}$$

Table II

Entropy data for metal redox reaction of Co(II)TPP in several solvent systems.

Solvent	ΔS, eu	
	Co(II) $\rightleftarrows$ Co(I) [a]	Co(II) $\rightleftarrows$ Co(III) [b]
Butyronitrile	0.0 ± 2	(d)
DMSO	6.2 ± 0.9	-35.0 ± 1.6
Py-DMSO [c]	16.2 ± 2.4	-43.5 ± 1.6

(a) [Co(II)TPP]° + e $\rightleftarrows$ [Co(I)TPP]⁻
(b) [Co(II)TPP]° $\rightleftarrows$ [Co(III)TPP]⁺ + e
(c) 1.13 M pyridine in DMSO
(d) Ill defined electrode reaction

Table III

Thermodynamic data for addition of pyridine to cobalt(II)tetra-phenylporphrin according to the reaction:

Co(II)TPP + Py $\rightleftarrows$ Co(II)TPP·Py

Solvent	$\Delta G_{298°}$ kcal/mole	ΔH kcal/mole	ΔS [a] eu	ΔS [b] eu
DMSO	-1.50±0.06	-4.5±0.6	-10.0±1.7	-10.0±2.6
Toluene [c]	-3.66±0.05	-8.5±1	-16±4	————

(a) Obtained from intercept of log K vs $\frac{1}{T}$ plot.
(b) Obtained from Hess's Law calculation involving electron transfer reactions of Co(II)TPP in the presence and absence of pyridine
(c) F. A. Walker, J. Amer. Chem. Soc., **95**, 1150 (1973)

for which ΔS = -10.0 ± 2.6 eu. This is listed in Table III.

In contrast to the invariance of half wave potentials for Co(II)TPP reduction (Reaction 3), oxidation to yield Co(III) (Reaction 4) is markedly dependent on solvent(22,26,27). Changing from a relatively low dielectric constant solvent such as benzonitrile to a solvent of higher dielectric constant such as DMSO produces a cathodic shift of over 350 mV.(22) This effect is even more dramatic on going from benzonitrile to pyridine where half wave potentials will shift by more than 600 mV on changing solvents.

The ease of Co(II) oxidation has been shown to parallel the coordinating ability of the solvent, indicating a stabilization of cobalt(III) with axial ligands.(22) The entropy data of Table II bear this out. The negative entropy changes are large for oxidation of Co(II) to Co(III) in both DMSO and DMSO-pyridine mixtures, and can best be attributed to addition of one or more solvent molecules. This has been postulated by Davis(26) and Manassen(27) and was confirmed in our laboratory by monitoring the half wave potentials for reaction 3 during a DMSO titration of Co(II)TPP in butyronitrile. From 0 to 10^{-2} M DMSO the oxidation potentials shifted only slightly. Above 10^{-2} M and up to 1M DMSO the observed shift was -68mV/log(DMSO). Above 1M DMSO this shift increased to -116mV/log(DMSO). Based on these results, the following formulation is given for complexation of Co(III) by DMSO:

$$[Co(II)TPP]^{\circ} + 2\ DMSO \underset{\leftarrow}{\overset{\rightarrow}{}} [Co(III)TPP\cdot DMSO_2]^+ + e \qquad (7)$$

for which a ΔS = -35.0 ± 1.6 eu is calculated.

Thermodynamics of Co(II)TPP Complexation with Pyridine in DMSO. Utilization of equation 2 permits calculation of both the stability constant and formula for the binding of Co(II)TPP by pyridine in DMSO. Electrochemically followed titrations of CoTPP with pyridine show that one ligand is bound to Co(II), while zero are bound to Co(I).(26) At 298°K a ΔG = -1.50±0.06 Kcal/mole was calculated for the mono pyridine complex. The temperature was varied, and from the resulting Van't Hoff plot in Figure 3, a ΔH = -4.5 ± 0.6 kcal/mole and ΔS = -10.0 ± 1.7 eu calculated for pyridine addition according to reaction 6. (These thermodynamic values are listed in Table III.) The ΔS agrees, within experimental error, with the ΔS = -10.0 ± 2.6 eu calculated from combination of reactions 3 and 5. It is also within the range of ΔS calculated in toluene by Walker.(11) In contrast, however, the enthalpy change, ΔH, is smaller by 4.0 kcal/mole from that obtained in toluene and would account for the large differences in ΔG between solvents.

Electrode Reaction of FeTPPCl. It is of some interest to draw parallels between the oxidations of Co(II)TPP and Fe(II)TPP.

Iron tetraphenylporphyrin will undergo two electrode reactions at the central metal,

$$[Fe(II)TPP]^\circ + e \rightleftarrows [Fe(I)TPP]^- \qquad (8)$$

and

$$[Fe(II)TPP]^\circ \rightleftarrows [Fe(III)TPP]^+ + e \qquad (9)$$

Both reactions, and especially reaction 9, are highly dependent on the coordinating ability of the solvent. For reaction 9 the half wave potentials are observed to shift anodically by up to 200 mV on going from a non-coordinating solvent such as methylene chloride or butyronitrile to a coordinating solvent such as DMF, DMA or DMSO.(28) This has been accounted for by an increased stability of $\overline{Fe(II)}$TPP by complexation with the solvent.

Since iron(I) does not complex with axial ligands at room temperature it can be predicted that reduction of Fe(II)TPP should be accompanied by a positive entropy change due primarily to ligand dissociation. This is exactly what is observed in Table IV.

Table IV

Entropy data for metal redox reactions of Fe(II)TPP in several solvent systems.

Solvent	ΔS, eu	
	Fe(II) $\rightleftarrows$ Fe(I) [a]	Fe(II) $\rightleftarrows$ Fe(III) [b]
Butyronitrile	7.5 ± 0.5	-2.5 ± 1.0
DMF	12.6 ± 1.0	-7.6 ± 0.7
DMSO	24.6 ± 1.0	3.5 ± 0.2
Py-Butyronitrile [c]	27.2 ± 2.0	-0.6 ± 0.5
Py-DMF [c]	29.5 ± 1.8	-0.5 ± 0.5
Py-DMSO [c]	39.4 ± 2.3	15.3 ± 3.2

(a) $[Fe(II)TPP]^\circ + e \rightleftarrows [Fe(I)TPP]^-$
(b) $[Fe(II)TPP]^\circ \rightleftarrows [Fe(III)TPP]^+ + e$
(c) 2.06 $\underline{M}$ pyridine

In DMF-pyridine solutions the electrode reaction is

$$[Fe(II)TPP \cdot Py_2]° + e \rightleftarrows [Fe(I)TPP]^- + 2Py \qquad (10)$$

for which $\Delta S = 29.5 \pm 1.8$ eu. Reversing the sign on reaction 10 and adding to reaction 11 in DMF

$$[Fe(II)TPP \cdot DMF_x]° + e \rightleftarrows [Fe(I)TPP]^- + xDMF \qquad (11)$$

yields the overall reaction 12 for formation of a bis pyridine adduct.

$$[Fe(II)TPP \cdot DMF_x]° + 2Py \rightleftarrows [Fe(II)TPP \cdot Py_2]° + xDMF \qquad (12)$$

for which $\Delta S = -16.9 \pm 2.0$ eu. Entropy calculations for this reaction in DMF as well as in DMSO and butyronitrile are listed in Table V.

Table V

Thermodynamic data for additon of pyridine to iron(II)tetraphenylporphyrin according to the reaction:

$$Fe(II)TPP + 2 Py \rightleftarrows Fe(II)TPP \cdot Py_2$$

Solvent	$\Delta G_{298°}$ kcal/mole	ΔH kcal/mole	$\Delta S^{(a)}$ eu	$\Delta S^{(b)}$ eu
DMSO	-3.6±0.2	-8.8±0.5	-17.4±1.2	-14.8±2.5
DMF	-7.3±0.2	-13.0±0.8	-19.2±2.0	-16.9±2.0
Butyronitrile	-8.9±0.1	-15.6±0.4	-22.5±1.8	-19.7±2.1

(a) Obtained from intercept of log K vs $\frac{1}{T}$ plot.
(b) Obtained from Hess's law calculation involving electron transfer reaction of Fe(II)TPP in the presence and absence of pyridine.

A typical plot of $dE_{\frac{1}{2}}/dT$ for reactions 10 and 11 is shown in Figure 4 and ΔS calculated from these plots is listed in Table V for several solvents. All entropy changes for Fe(II) complexation in Table V are within the range reported for other iron(II) porphyrins, but, as has been pointed out, ΔS (as well as ΔH) for binding of 2 pyridine molecules is extremely solvent depenent.(13)

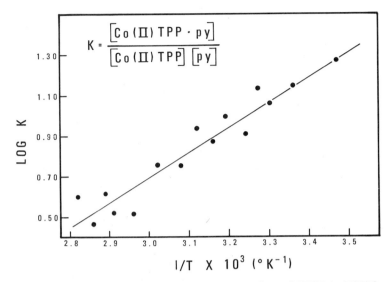

Figure 3. Van't Hoff plot for pyridine binding by Co(II)TPP in DMSO, 0.1M TBAP

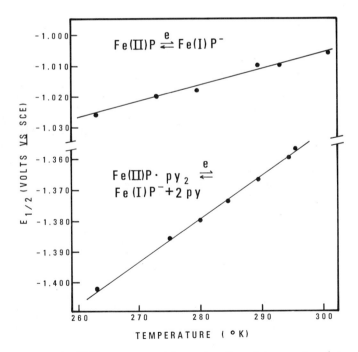

Figure 4. Half wave potential as a function of temperature for Reactions 10 and 11

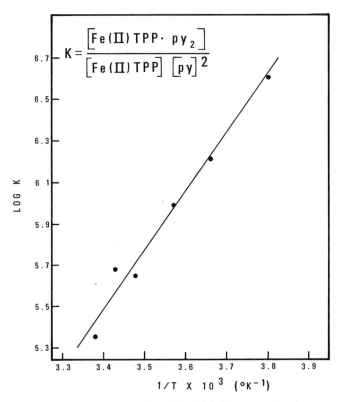

Figure 5. Van't Hoff plot for pyridine binding by Fe(II)TPP in DMF, 0.1M TBAP

In contrast to Co(II)TPP, ΔS for oxidation of Fe(II) to
Fe(III) (see Table IV) is quite small in pyridine-DMF and pyri-
dine-butyronitrile mixtures where both iron(II) and iron(III)
form bis pyridine adducts. ΔS can, in this case, be assigned
entirely to the electron transfer step without contribution from
changes in axial ligand coordination. The positive ΔS = 15.3 ±
3.2 eu in Py-DMSO implies a reduction in coordination on going
from Fe(II) to Fe(III). Likewise, the small entropy changes in
pyridine free solutions imply an identical coordination number of
both the oxidized and the reduced forms of the complex.

In order to corroborate the Hess's law determination for
complexation with iron(II), as has been done for Co(II)TPP, we
have calculated ΔS directly from the temperature dependence of
the measured stability constants. Calculations utilizing equation 2 gave a K = 2.2 x 10^5 at 298° for FeTPP·Py$_2$ in DMF and a
similar K = 3.2 x 10^6 in butyronitrile. The temperature was then
varied and from the slope of log K vs 1/T (Figure 5), values of
ΔH and ΔS were calculated, and listed in Table V. No difference
appears to exist between entropy changes calculated by the different methods.

Acknowledgements. The support of Research Corporation is great-
fully acknowledged.

Literature Cited

1. J.-H. Fuhrhop in "Structure and Bonding", Vol. 18, J. D.
   Dunitz, Ed., Springer-Verlag, New York, 1974.
2. K. M. Smith "Porphyrins and Metalloporphyrins", Elsevier
   Scientific Publishing Co., New York, N. Y., 1975, chapter
   14.
3. M. J. Carter, D. P. Rillema and F. Basolo, J. Amer. Chem.
   Soc., 96, 392 (1974).
4. H. C. Stynes and J. A. Ibers, J. Amer. Chem. Soc., 94, 1559
   (1972).
5. F. A. Walker, J. Amer. Chem. Soc., 95, 1154 (1973).
6. D. V. Stynes, H. Stynes, J. A. Ibers and B. R. James,
   J. Amer. Chem. Soc., 95, 1142 (1973).
7. C. J. Weschler, D. C. Anderson and F. Basolo, J. Amer.
   Chem. Soc., 97, 6707 (1975).
8. C. J. Weschler, D. C. Anderson and F. Basolo, J. Amer.
   Chem. Soc., 96, 5599 (1974).
9. C. K. Chang and T. G. Traylor, Proc. Nat. Acad. Sci.,
   U.S.A., 72, 1177 (1975).
10. C. J. Weschler, B. M. Hoffman and F. Basolo, J. Amer.
    Chem. Soc., 97, 5278 (1975).
11. F. A. Walker, J. Amer. Chem. Soc., 95, 1150 (1973).
12. D. V. Stynes, H. C. Stynes, B. R. James and J. A. Ibers,
    J. Amer. Chem. Soc., 95, 1797 (1973).

13. S. J. Cole, G. C. Curthoys and E. A. Magnusson, J. Amer. Chem. Soc., 92, 2991 (1970), ibid, 92, 2153 (1971).

14. W. J. Moore, "Physical Chemistry", Prentice Hall Inc., Englewood Cliffs, N.J., 1972.

15. I. M. Kolthoff and J.J. Lingane, "Polarography", Vol. 1, Interscience Publishers, New York, N. Y., 1952, chapter 12.

16. R. H. Felton and H. Linschitz, J. Amer. Chem. Soc., 88, 1113 (1966).

17. J. Fajer, D. C. Borg, A. Forman, D. Dolphin and R. H. Felton, J. Amer. Chem. Soc., 92, 3451 (1970).

18. R. P. Van Duyne and C. N. Reilley, Anal. Chem., 44, 142 (1972).

19. D. J. Quimby and F. R. Longo, J. Amer. Chem. Soc., 97, 5111 (1975).

20. D. Dolphin, R. H. Felton, D. C. Borg and J. Fajer, J. Amer. Chem. Soc., 92, 743 (1970).

21. K. A. Zachariasse and D. G. Whitten, Chem. Phys. Letters, 22, 527 (1973).

22. F. A. Walker, D. Beroiz, and K. M. Kadish, J. Amer. Chem. Soc., 98 3484 (1976).

23. F. A. Walker, private communication.

24. D. Lexa and J. M. Lhoste, Experimentia Suppl. 18, 395 (1971).

25. D. Beroiz and K. M. Kadish and L. Bottomley, manuscript in preparation.

26. L. A. Truxillo and D. G. Davis, Anal. Chem., 47, 2260 (1975).

27. J. Manassen, Isr. J. Chem., 12, 1059 (1974).

28. K. M. Kadish, M. M. Morrison, L. A. Constant, L. Dickens and D. G. Davis, J. Amer. Chem. Soc., in press.

Electrochemical Investigations of the Redox Properties of a N-Bridged Dimer, μ-Nitrido-bis[α,β,γ,δ-tetraphenylporphyriniron], in Nonaqueous Media

K. M. KADISH* and J. S. CHENG
Department of Chemistry, California State University, Fullerton, Calif. 92634

I. A. COHEN and D. SUMMERVILLE
Brooklyn College of the City University of New York, Brooklyn, N. Y. 11210

Investigations of iron porphyrin redox properties in nonaqueous media have led, in recent years, to the characterization of several previously unreported iron oxidation state complexes. Iron(I) porphyrins have been characterized both chemically(2) and electrochemically(3,4) while electrochemical studies of porphyrin π cation radicals and dications have led to the proposed assignment of quadrivalent heme iron (5,6). This latter oxidation state was generated by electrooxidation of either monomeric or dimeric iron(III) complexes of octaethyl or tetraphenylporphyrin(5,6). With the μ-oxo dimers, only one of the two iron atoms was oxidized to yield a mixed oxidation state Fe(III)-Fe(IV) dimer. Electroreduction of μ-oxo-bis-[tetraphenylporphinato-iron(III)], (FeTPP)₂O, in DMF has also been shown to yield a mixed Fe(III)-Fe(II) complex, which was characterized by e.s.r. at low temperature(4) before dissociation to [Fe(I)TPP]⁻.

Recently, synthesis of the first stable nonintegral or mixed oxidation state iron porphyrin dimer was reported by Summerville and Cohen(7). This is a nitrogen-bridged species, μ-nitrido-bis-[α,β,γ,δ-tetraphenylporphinatoiron], written as (FeTPP)₂N, and similar to, but not isoelectronic with, (FeTPP)₂O. In neutral (FeTPP)₂O, the formal oxidation state on both irons is +3. In (FeTPP)₂N, however, the extra negative charge on the bridging atom N³⁻, when compared to O²⁻, leads to the average iron oxidation state +3½. The neutral 17 valence electron nitrido complex is thus isoelectronic with the cationic species [(FeTPP)₂O]⁺ characterized by Felton(5,6) while the reduced nitrido complex, [(FeTPP)₂N]⁻, is isoelectronic with the well characterized 18 valence electron (FeTPP)₂O.(4,9-12)

Comparisons of the physical properties(7) and X-Ray structure(8) of the neutral (FeTPP)₂N and (FeTPP)₂O have recently been made. The most outstanding difference between these two systems is the extent of antiferromagnetic coupling between the iron atoms accross the bridge. Whereas (FeTPP)₂N is a completely coupled dimer, (FeTPP)₂O is only weakly magnetically coupled. Because of the relationship between spin coupling and bridge

*Present address: Department of Chemistry, University of Houston, Houston, Texas 77004

mediated electron transfer and its relevance to biological electron transport in cytochrome oxidase, we wish to compare a series
of isoelectronic hemin dimers utilizing different bridging groups.
In this paper, we report the electron transfer properties of
μ-nitrido-bis [$\alpha,\beta,\gamma,\delta$-tetraphenylporphinatoiron] in methylene
chloride.

Experimental

Chemicals. All solvents and chemicals were reagent grade
and were used without further purification. The supporting
electrolyte, tetrabutylammonium perchlorate (TBAP) was recrystallized from absolute methanol and was dried at reduced pressure
over P_4O_{10}. (FeTPP)$_2$O was purchased from Strem Chemical Inc.
(Danvers, Mass.) and was used as received. (FeTPP)$_2$N was synthesized from TPPFeN$_3$ as described by the method of Summerville and
Cohen (7).

Electrochemical Measurements. All polarographic measurements
were made on a PAR Model 174 Polarographic Analyzer utilizing a
three electrode system. The working electrode and counter electrode were platinum and a commercial calomel electrode was utilized as the reference electrode. This was separated from the bulk
of the solution by a bridge filled with the same solvent and
supporting electrolyte. Porphyrin concentrations were between
10^{-3} and 10^{-4} M. The overall number of electrons (faradays per
mole of iron monomer) was determined by controlled potential
coulometry utilizing a PAR Model 173 Potentiostat. Electronic
integration of the current-time curve was achieved using a PAR
Model 179 integrator in conjunction with the Model 173 Potentiostat. The Coulometric cell was similar to that used for cyclic
voltammetry. A large coiled platinum wire served as the anode
and was separated from the cathodic compartment by means of a
fritted disk. A platinum mesh electrode was used as the cathode
and a saturated calomel electrode was the reference electrode.
Stirring of the solution was achieved by means of a magnetic
stirring bar. Dearation of the solution was performed before
commencing the experiment and a stream of high purity argon was
passed above the solution throughout the experiment. All experiments were carried out in a controlled temperature room of 20±0.5°
and all potentials are reported vs the saturated calomel electrode
(SCE).

Optical Spectroscopy. The electrolysis of (FeTPP)$_2$N was
followed optically using a Cary 15 Spectrophotometer. A specially
constructed quartz flow cell of path length 0.90 cm was used,
which was attached to the electrolysis cell and could be removed
for insertion into the Cary 15. For the neutral species, quartz
spectrophotometric cells of path length 1.00 and 0.10 cm were
also utilized.

Results

Cyclic Voltammetry and Differential Pulse Polarography.   The electrochemical reduction of $(FeTPP)_2N$ proceeds in several discrete steps without destroying the porphyrin ring.   In Figure la is shown a cyclic voltammogram of $(FeTPP)_2N$ obtained in CH_2Cl_2. Also shown in this figure are cyclic voltammograms of $(FeTPP)_2O$ and FeTPPCl in the same solvent.   For comparison, differential pulse polarograms are shown overlapping the cyclic voltammograms. The advantage of differential pulse polarography is that peak current height may be accurately measured for closely overlapping reactions which cannot be analyzed by cyclic voltammetry.   In addition, half wave potentials are readily obtainable from the position of the peak $(E_{\frac{1}{2}} \approx E_p$ at small modulation amplitudes (13)). Half wave potentials in CH_2Cl_2 were identical by each method and are summarized in Table 1.

Table I

Half Wave Potentials for Electrooxidation-Reduction of Several Similar Porphyrins at a Platinum Electrode in CH_2Cl_2, 0.1M TBAP

Half Wave Potential (volts vs. SCE)

| Compound | Reaction[a] | | | | |
	(4)	(3)	(2)	(1)	(5)
$(FeTPP)_2N$	1.76	1.51	1.15	0.15	-1.21
$(FeTPP)_2O$	--	1.45	1.09	0.84	-1.17
FeTPPCl	--	1.63	1.42	1.14	-0.32

(a)   See Figure 1 and text for identification of each peak

In order to investigate each reaction process of $(FeTPP)_2N$ separately and to determine the existence of any chemical reactions coupled to the electron transfer, cyclic voltammograms were taken over various sweep ranges.   The potential was initially set at 0.4 V and scanned in successively larger increments first up to -1.6 V in a cathodic direction and then up to +1.9 V in an anodic direction.   For either single or multiple scans between the range of +1.9 V and -1.6 V, a diffusion controlled reduction and reoxidation was invariably observed at 0.15 V in methylene chloride.   This is labeled peak 1 in Figure la and can be assigned to

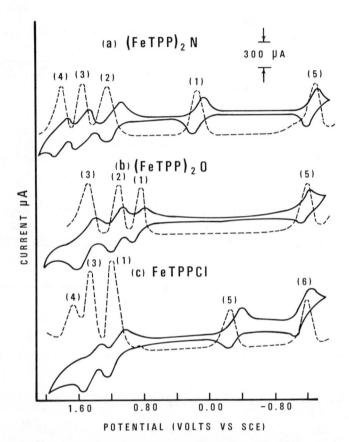

Figure 1. Cyclic voltammograms obtained at 100 mV/sec on a platinum electrode (——) and differential pulse polarograms at 2 mV/sec, modulation amplitude 25 mV/sec on a platinum electrode (– – –) for (a) (FeTPP)₂N; (b) (FeTPP)₂O; and (c) FeTPPCl in CH₂Cl₂, 0.1M TBAP

the transition:

$$[\text{TPPFe(III)-N-Fe(IV)TPP}]^{o} + e- \underset{\leftarrow}{\rightarrow} [\text{TPPFe(III)-N-Fe(III)TPP}]^{-}$$

This same reaction was observed at a similar potential in benzo-nitrile (Table II) but in pyridine this reduction peak was shifted cathodically in potential to -0.26 V. This was not due to the reduction of FeTPPCl·py$_2$ which is reduced at 0.18 and -1.38 V in neat pyridine (14). No further reduction was observed in any range of scans up to cathodic potentials of at least -1.1 V.

Table II

Half Wave Potentials for Oxidation-Reduction of (FeTPP)$_2$N in several solvents, 0.1M TBAP

Solvent	Half Wave Potential (volts vs SCE)				
			Reaction[a]		
	(4)	(3)	(2)	(1)	(5)
CH$_2$Cl$_2$	1.76	1.51	1.15	0.15	-1.21
Benzonitrile	(b)	1.22	1.06	0.16	-1.17
Pyridine	(b)	(b)	(b)	-0.26	-1.15

(a)   See Figure 1a for CH$_2$Cl$_2$

(b)   Beyond potential range of solvent

When the scan was extended to -1.21 V a second reduction peak (5 of Figure 1a) was obtained. This is at almost an identical potential to the first reduction of (FeTPP)$_2$O at -1.17 V in CH$_2$Cl$_2$ and the second reduction of FeTPPCl at -1.08 V. No further reduction was observed up to the solvent limit of -1.7 V.

Currents and half wave potentials for peaks 1,2 and 5 of (FeTPP)$_2$N are listed in Table III for slow scan rates. As seen from this table the invariance of peak current with the square root of the scan rate, as well as the constant half wave potential, indicates clearly a diffusion controlled electron transfer for each reaction. Peak currents from differential pulse polarograms were constant for each reaction of (FeTPP)$_2$N, and polarographic wave analysis gave a slope of 63mV, indicating again a single electron transfer step.

Table III

Scan Rate Dependence of Half Wave Potential and Peak Current for Three Electrode Reactions of 8.14 $\times 10^{-4}$ M $(FeTPP)_2N$ in 0.1M TBAP in Methylene Chloride.

Reaction(a)

Scan Rate	(1)			(5)			(2)		
(mV/sec)	$E_{\frac{1}{2}}$ (Volt)	$i_{p,c}$ (μA)	$\dfrac{i_{p,c}/v^{\frac{1}{2}}}{μA \cdot sec^{\frac{1}{2}}}$ $mV^{\frac{1}{2}}$	$E_{\frac{1}{2}}$ (Volt)	$i_{p,c}$ (μA)	$\dfrac{i_{p,c}/v^{\frac{1}{2}}}{μA \cdot sec^{\frac{1}{2}}}$ $mV^{\frac{1}{2}}$	$E_{\frac{1}{2}}$ (Volt)	$i_{p,a}$ (μA)	$\dfrac{i_{p,a}/v^{\frac{1}{2}}}{μA \cdot sec^{\frac{1}{2}}}$ $mV^{\frac{1}{2}}$
10	0.147	6.0	1.90	–	–	–	–	–	–
20	0.148	8.4	1.88	-1.211	7.5	1.68	+1.149	8.4	1.88
50	0.148	13.2	1.88	-1.213	12.2	1.73	+1.147	12.4	1.75
100	0.148	19.2	1.92	-1.210	17.5	1.75	+1.144	17.6	1.76
200	0.148	26.8	1.90	-1.210	23.1	1.63	+1.144	24.4	1.73
500	0.146	40.8	1.82	-1.216	35.0	1.57	–	–	–

(a) See figure 1a.

Controlled Potential Electrolysis. In order to identify the products of each electrode reaction, the neutral species was both electroreduced and electrooxidized at controlled potential and the number of coulombs recorded by integration of the resulting current-time curve. The voltammogram before electrolysis is shown in Figure 1a. The potential was then set at -0.6 V. This is on a plateau of the first reduction wave but 600mV anodic of the second reduction. Controlled potential electrolysis was complete after 10 minutes at this potential and yielded an n = 0.50 electrons/monomeric unit. The potential was then set to -1.4 V and (FeTPP)$_2$N was reduced at a controlled potential. The 60mV separation of the cathodic and anodic peak on the cyclic voltammogram (Figure 1) as well as polarographic wave analysis indicate a reversible one electron transfer reaction at =-1.21 V. However, controlled potential reduction did not yield current-time curves indicative of a single electron transfer process but gave evidence of a chemical reaction coupled to the electron transfer step. Calculations of n for this step were not reproducible, suggesting that the dimer was cleaved during electroreduction. Electrooxidations were performed at +1.13, +1.65 and +1.90 V. In each case a well defined integrated current-time curve was obtained, with a n=0.5, 0.99 and 1.35 electrons/monomeric unit as the solution color changed from brown to green. Values of n at each potential are summarized in Table IV.

Table IV

Controlled Potential Electrolysis and Coulometry of (FeTPP)$_2$N in CH$_2$Cl$_2$

Potential(v)	Electrons Transfered [a]	Reaction
-0.60 [b]	0.50	$[Fe(IV)-N-Fe(III)]+e^- \rightleftharpoons [Fe(III)-N-Fe(III)]^{-1}$
1.30 [c]	0.50	$[Fe(IV)-N-Fe(III)] \rightleftharpoons [Fe(IV)-N-Fe(III)]^{+1}+e^-$
1.65 [d]	0.99	$[Fe(IV)-N-Fe(III)] \rightleftharpoons [Fe(IV)-N-Fe(III)]^{+2}+2e^-$
1.90 [e]	1.35	$[Fe(IV)-N-Fe(III)] \rightleftharpoons [Fe(IV)-N-Fe(III)]^{+3}+3e$

(a) Values given as electrons/monomeric unit
(b) 750 mV more cathodic than Reaction (1)
(c) 150 mV more anodic than Reaction (2)
(d) 140 mV more anodic than Reaction (3)
(e) 140 mV more anodic than Reaction (4)

Optical Spectra. Before electrolysis the spectrum of
(FeTPP)$_2$N consisted of a split Soret Band in the UV region (λ =
408 and 385 nm) and two weaker bands in the visible region (λ =
625 and 532 nm). A methylene chloride solution of (FeTPP)$_2$N was
stable in air and showed no change in spectral properties for
several days. Values of these molar absorptivities are listed in
Table V, Group I. For comparison we have also listed λ and ε for
the spectrum of isoelectronic [(FeTPP)$_2$O]$^+$ obtained by Felton.(6)

Optical spectra of the singly reduced species (Table 5,
Group 2) show the features of an unsplit Soret band (λ = 396 nm)
which is of approximately the same molar absorptivity as that for
the isoelectronic (FeTPP)$_2$O but shifted toward the blue. There
is no structured absorption of [(FeTPP)$_2$N]$^-$ in the 500-700 nm
region. This species was stable in the presence of O$_2$ for over
24 hours and was unchanged from that obtained under an insert
argon atmosphere. Reoxidation, however, did not yield the original starting spectrum. The spectra of (FeTPP)$_2$N, [(FeTPP)$_2$N]$^-$
and (FeTPP)$_2$O are displayed in Figures 2 and 3.

As previously mentioned, reduction at -1.4 V did not proceed
in a single step and yielded, in all cases, an ultimate monomeric
product after several hours. Products of the oxidized species
were also unstable in the time interval of the experiment and the
spectra resembled those reported for FeTPP^{+2}.(6)

Discussion of Results

Based on the data, the electrooxidation-reduction of
(FeTPP)$_2$N can be accounted for by the following mechanism:

$$[TPPFe^{III\frac{1}{2}}-N-Fe^{III\frac{1}{2}} TPP]^0$$

$-e \quad E_{\frac{1}{2}} = 1.15 \text{ V} \qquad +e \quad E_{\frac{1}{2}} = 0.15 \text{ V}$

$[TPPFe^{III}-N-Fe^{IV}TPP]^+ \qquad\qquad [TPPFe^{III}-N-Fe^{III}TPP]^{-1}$

$-e \quad E_{\frac{1}{2}} \quad 1.51 \text{ V} \qquad\qquad +e \quad E_{\frac{1}{2}} = -1.21 \text{ V}$

$[TPPFe^{III}-N-Fe^{IV}TPP]^{+2} \qquad\qquad [TPPFe^{III}-N-Fe^{III}TPP]^{-2}$

$-e \quad E_{\frac{1}{2}} = 1.76 \text{ V} \qquad\qquad\qquad$ further reduction products

$[TPPFe^{III}-N-Fe^{IV}TPP]^{+3}$

The initial reduction of (FeTPP)$_2$N is quite facile and
yields initially an iron(III) dimer isoelectronic with (FeTPP)$_2$O.
Further reduction of the dimer occurs at -1.21 V by a single
electron transfer step and yields a product assigned not as a
mixed Fe(III), Fe(II) dimer, but rather as a dimeric iron(III)
anion radical.

Table V

Absorbance Maxima (nm) and Molar Absorptivities (l·mole⁻¹·cm⁻¹) for Several Iron Tetraphenylporphyrin Complexes in Methylene Chloride. Values of Molar Absorptivities Are Calculated Per Mole of Iron.

Group	Compound					Reference
I	[Fe(IV)-N-Fe(III)]°	λ	532	408	385	This work
		ε x 10⁻⁴	0.86	10.7	11.3	
	[Fe(IV)-O-Fe(III)]⁺	λ	(a)	408	327	Ref. 6
		ε x 10⁻⁴	—	8.50	3.60	
II	[Fe(III)-N-Fe(III)]⁻	λ	(a)	396	—	This work
		ε x 10⁻⁴	—	9.40	—	
	[Fe(III)-O-Fe(III)]°	λ	613	562	408	This work
		ε x 10⁻⁴	0.43	0.71	10.0	
III	Fe(III)TPPCl	λ	510	417	378	Ref. 6
		ε x 10⁻⁴	1.34	11.3	6.34	

(a) No spectral detail in visible region.

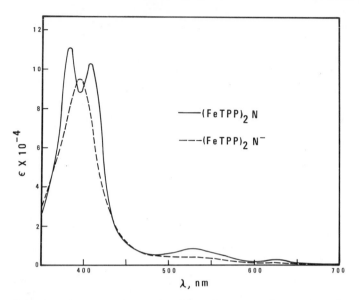

Figure 2.   Spectra of (FeTPP)₂N first reduction product in CH₂Cl₂, 0.1M TBAP. (FeTPP)₂N before electrolysis (——); after controlled potential reduction at −0.60 V to yield [(FeTPP)₂N]⁻ (−−−). The molar absorptivity is per monomeric iron.

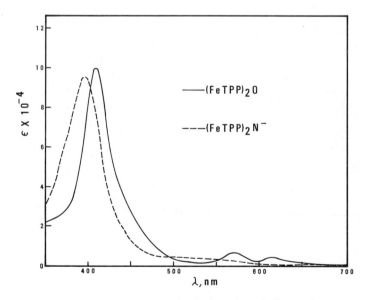

Figure 3.   Spectra of [(FeTPP)₂N]⁻ (−−−) and the isoelectronic (FeTPP)₂O (——) in CH₂Cl₂, 0.1M TBAP. The molar absorptivity is per monomeric iron.

Oxidation of $(FeTPP)_2N$ occurs in three well defined single electron transfer steps (see figure 1). We have chosen to assign the reaction as occuring at the porphyrin ring to yield a cation radical rather than at the central metal to yield an Fe(IV), Fe(IV) dimer. The reasons for these assignments are based only on potentials of the redox reactions and will be discussed in the following sections.

The starting material $(FeTPP)_2N$ has been well characterized from Mössbauer data and magnetic susceptibility measurements. At room temperature the Mössbauer spectrum is sharp and symmetrical and at 80°K shows only a slight asymmetry. Magnetic susceptibility measurements between 80 K and 300 K indicate a simple paramagnetic species with $\mu = 2.04$ BM per $(FeTPP)_2N$. Since the Mössbauer data presents a temperature independent doublet this would indicate that either the two iron atoms are in the same oxidation state $(3\frac{1}{2})$ and environment, or alternately there exists the occurance of very rapid exchange between the two iron centers $(<10^{-7}$ sec) as shown below.

$$[TPPFe(III)-N-Fe(IV)TPP] \rightleftarrows [TPPFe(IV)-N-Fe(III)TPP]$$

The equivalence of the two iron atoms has been confirmed by the X-ray structural results (8) and an average oxidation state greater then +III is consistent with the observed Mössbauer isomer shift for $(FeTPP)_2N$ (7). Thus the formulation of iron $(3\frac{1}{2})$ is prefered.

On the other hand, the isoelectric cation $[(FeTPP)_2O]^+$ has been shown to have a temperature dependent moment similar to $(FeTPP)_2O$, and might be represented as having two non-equivalent iron atoms [Fe(III) and Fe(IV)] which do not exchange rapidly, or two equivalent iron atoms which are not strongly magnetically coupled. Thus, it is of some interest to compare both oxidation and reduction potentials for reactions of two isoelectronic iron(IV) dimers, $[(FeTPP)_2O]^+$ and $(FeTPP)_2N$ (Reaction 2), and $(FeTPP)_2O$ and $[(FeTPP)_2N]^-$ (Reaction 5).

As seen from Table I, the half wave potentials for oxidation of an Fe(IV), Fe(III) dimer (Reaction 2) are about equal for the isoelectronic $(FeTPP)_2N$ and $[(FeTPP)_2O]^+$. The former is oxidized at 1.15 V while the latter reaction occurs at 1.09 V. This 60 mV difference is not large and the half wave potentials do not seem to be greatly influenced by the extra positive charge on $[(FeTPP)_2O]^+$. Likewise, the isoelectronic $[(FeTPP)_2N]^-$ and $(FeTPP)_2O$ have almost identical reduction potentials with the former being measured at -1.21 V and the latter at -1.17 V. Again, this 40 mV difference is not large and can be accounted for by the extra negative charge on $[(FeTPP)_2N]^-$ which retards slightly the reduction when compared to the neutral $(FeTPP)_2O$.

In marked contrast, however, the potentials for reduction at the central metal of the Fe(III)-Fe(IV) dimer are extremely sensitive to the bridging atom. The range of stability of the

Fe(III)-Fe(IV) dimers $\underline{vs}$ the Fe(III)-Fe(II) dimers indicates that the N bridge stabilizes the Fe(III), Fe(IV) oxidation state by almost 700mV when compared to the O bridge. This is consistant with the properties of N^{-3} vs O^{-2} as the bridging atom. The greater negative charge on the N^{-3} bridge, the larger size of N^{-3} and the greater magnitude of π bonding with N^{-3}, all favor the higher oxidation state of the iron atoms in the N bridged dimer when compared to the O bridged dimer. X-Ray data shows that the Fe-N distance is 1.6605_0A in (FeTPP)$_2$N which is considerably shorter than the 1.763 A in (FeTPP)$_2$O. This is consistant with the infrared data which indicates a greater degree of π charge delocalization for Fe-N-Fe vs Fe-O-Fe.

However, despite this difference between bridging atoms, the potential difference, Δ, between reactions (2) and (5) are about equal with Δ = 2.36 V for (FeTPP)$_2$N and Δ = 2.26 V for (FeTPP)$_2$O. This is within the range of Δ = 2.25±0.15 V which has been reported for a series of cation and anion radical reactions of octaethyl and tetraphenylporphyrin(14,15) complexes independent of changes in central metal oxidation state. Accordingly, we would like to propose that the electrode reactions (2) and (5) correspond to the formation of the cation radical and anion radical, respectively. The assignment of a cation radical product has been previously reported (5,6) while that of an anion radical product differs from an earlier characterization of [(FeTPP)$_2$O]$^-$ in DMF (4). In this solvent, an Fe(III), Fe(II) dimer was observed. Further studies aimed at characterization of the reduction products of (FeTPP)$_2$O and (FeTPP)$_2$N in several solvents are now in progress.

Literature Cited

1. (a) California State University, Fullerton (b) Brooklyn College
2. Cohen, I.A., Ostfeld, D., and Lichtenstein, B., J. Amer. Chem. Soc., (1972) 94, 4522.
3. Lexa, D., Momenteau, M., and Mispelter, J., Biochim. Biophys. Acta., (1974), 338, 151.
4. Kadish, K. M., Larson, G., Lexa, D., and Momenteau, M., J. Amer. Chem. Soc., (1975), 97, 282.
5. Felton, R. H., Owen, G. S., Dolphin, D. and Fajer, T., J. Amer. Chem. Soc., (1971), 93, 6332.
6. Felton, R. H., Owen, G. S., Dolphin, D., Forman, A., Borg, D. C., and Fajer, T., Ann. N.Y. Acad. Sci., (1973), 206, 504.
7. Summerville, D. A. and Cohen, I. A., J. Amer. Chem. Soc., (1976), 98, 1747.
8. Scheidt, W. R., Summerville, D. A. and Cohen, I. A., J. Amer. Chem. Soc., (1976), 98, 6623.
9. Torrens, M. A., Straub, D. K. and Epstein, L. M., J. Amer. Chem. Soc., (1972), 94, 4160.

10. Cohen, I. A., J. Amer. Chem. Soc., (1969), 91, 1980.
11. Hoffman, A. B., Collins, D. M., Day, V. W., Fleischer, E. B., Srivastara, T. S. and Hoard, J. L., J. Amer. Chem. Soc., (1972), 94, 3620.
12. LaMar, G. N., Eaton, G. R., Holm, R. H. and Walker, F. A., J. Amer. Chem. Soc., (1972), 94, 3620.
13. J. H. Christie, J. Osteryoung and R. A. Osteryoung, Anal. Chem., (1973), 45, 210.
14. Morrison, M. and Kadish, K., unpublished results.
15. Fuhrhop, J. H., Kadish, R. M. and Davis, D. G., J. Amer. Chem. Soc., (1973), 95, 5140.
16. Kadish, K. M., Davis, D. G. and Fuhrhop, J. H., Angew. Chem. (Int. Edit.), (1972), 11, 1014.

6

Electrochemically Catalyzed Reduction of Nitrogenase Substrates by Binuclear Molybdenum(V) Complexes

FRANKLIN A. SCHULTZ, DEBRA A. LEDWITH, and LOUIS O. LEAZENBEE

Department of Chemistry, Florida Atlantic University, Boca Raton, Fla. 33431

A widely cited model for nitrogenase enzyme (1) is based on the binuclear di-μ-oxo-bridged Mo(V)-cysteine complex, $Na_2Mo_2O_4(Cys)_2$ (2). Solutions of this compound plus a chemical reducing agent ($NaBH_4$ or $Na_2S_2O_4$) catalytically reduce the enzyme substrates dinitrogen, acetylene, nitriles, and isonitriles in mildly alkaline (pH 7-12) aqueous media. The mechanism for these catalytic reductions (see Scheme I in ref. 3) is proposed to consist of the following sequence of reactions: dissociation of the binuclear Mo(V) complex, reduction to a monomeric Mo(IV) species, binding and reduction of substrate, and completion of the catalytic cycle by reduction of the oxidized catalyst with BH_4^-.

$$\frac{1}{2} Mo_2O_4(Cys)_2^{2-} \rightleftarrows Mo(V)-Cys \qquad (1)$$

$$Mo(V)-Cys + e^-(BH_4^-) \rightarrow Mo(IV)-Cys \qquad (2)$$

$$Mo(IV)-Cys + C_2H_2 \rightleftarrows Mo(IV)-Cys-C_2H_2 \qquad (3)$$

$$Mo(IV)-Cys-C_2H_2 + 2H^+ \rightarrow Mo(VI)-Cys + C_2H_4 \qquad (4)$$

$$Mo(VI)-Cys + 2e^-(BH_4^-) \rightarrow Mo(IV)-Cys \qquad (5)$$

Despite the success of this model in simulating many of the reactions of nitrogenase enzyme, relatively little is known about the oxidation-reduction chemistry of the binuclear $Mo_2O_4^{2+}$ center and the means by which an active catalyst is generated from this species. A major research objective in our laboratory has been to characterize the electrochemical behavior of binuclear molybdenum complexes. We have recently reported detailed electrode reaction mechanism studies of the principal nitrogenase model compound, $Na_2Mo_2O_4(Cys)_2$ (4), and its EDTA analog, $Na_2Mo_2O_4(EDTA)$ (5), in aqueous borate, phosphate, and ammonia buffers. This work is now being extended to a series of cysteine and EDTA complexes containing μ-oxo-μ-sulfido ($Mo_2O_3S^{2+}$) and

78

and di-μ-sulfido ($Mo_2O_2S_2^{2+}$) bridged Mo(V) cores (Figure 1) (6).
Not surprisingly, we also have found that electrochemical reduction
of $Na_2Mo_2O_4(Cys)_2$ and other oxo- and sulfido-bridged Mo(V)
species in the presence of acetylene leads to catalytic reduction
of this substrate (7). Aside from demonstrating that catalysis
can be initiated electrochemically, two objectives in this work
have been 1) to use information from electrode reaction mechanism
studies to help determine the nature and oxidation state of the
active catalyst, and 2) to use control of electrochemical
variables and solution conditions to provide insight to the
mechanism of catalysis. For the oxo- and sulfido-bridged Mo(V)
complexes features such as ligand and bridging atom significantly
influence the electrochemical and catalytic properties of the
binuclear center. The results of these studies provide a general
framework for understanding the mode of production of active
catalysts from binuclear Mo(V) complexes and the mechanism by
which these species catalyze the reduction of nitrogenase
substrates.

Electrochemistry of Oxo- and Sulfido-Bridged Complexes of Mo(V)

All of the oxo- and sulfido-bridged Mo(V)-cysteine and EDTA
complexes are reduced to binuclear Mo(III) products in a single,
diffusion-controlled step at ca. -1.1 to -1.3 V vs. SCE in
0.1 F $Na_2B_4O_7$. Some experimental results are shown in Figures 2
and 3 and Table I. Controlled potential coulometry and compara-
tive voltammetric and chronoamperometric current measurements
confirm that four electrons are transferred in the reduction of
each complex. The Mo(III)$_2$ electrode products are reoxidized to
Mo(V)$_2$ species at potentials 200-500 mV positive of the initial
reduction peak. The quasireversible character of this electron
transfer process is dependent upon both the structure of the
complex and the composition of the buffering medium.
Several important effects of ligand and bridge atom substitution are apparent from the electrochemical data. Replacement
of O by one or two bridging S atoms greatly increases the
reversibility of the Mo(V)$_2$/Mo(III)$_2$ electron transfer process,
as indicated by the decrease in ΔE_{peak}. Sulfur bridging atoms
also influence the stability of the dimeric Mo(III) electrode
products. The reverse peak currents shown in the cyclic voltammetric experiments in Figure 2 indicate that the sulfido-bridged
Mo(III)-cysteine products are less stable than the di-μ-oxo
analog, and that the rate of decomposition increases in the
order: $Mo_2O_4^{2+} < Mo_2O_3S^{2+} < Mo_2O_2S_2^{2+}$. All of the oxo- and
sulfido-bridged Mo(III)$_2$-EDTA products are stable on the time
scale of cyclic voltammetry (Figure 3). The $Mo_2O_4(EDTA)^{2-}$
complex can be carried through a complete coulometric reduction
and reoxidation cycle with its oxo-bridged structure intact (5).
However, changes in absorption spectra following reduction of
$Mo_2O_2S_2(EDTA)^{2-}$ indicate that its reduction product undergoes a

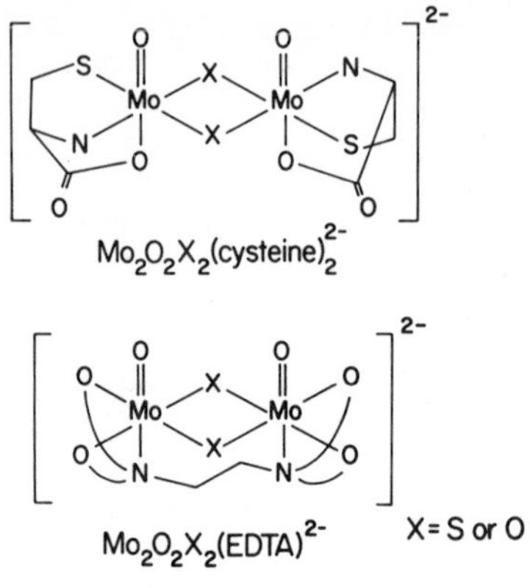

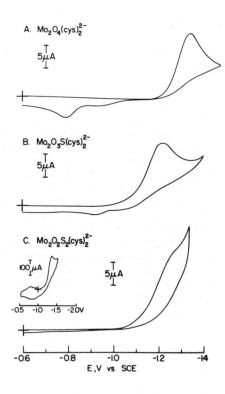

Mo
X
N
S
O
2-
$Mo_2O_2X_2(cysteine)_2^{2-}$
$Mo_2O_2X_2(EDTA)^{2-}$
X = S or O

Figure 1. Structures of the binuclear oxo- and sulfido-bridged molybdenum(V)–cysteine and EDTA complexes

A. $Mo_2O_4(cys)_2^{2-}$
5μA
B. $Mo_2O_3S(cys)_2^{2-}$
5μA
C. $Mo_2O_2S_2(cys)_2^{2-}$
100μA
5μA
-0.5 -1.0 -1.5 -2.0V
-0.6
-0.8
-1.0
-1.2
-1.4
E,V vs. SCE

Figure 2. Cyclic voltammetric curves for reduction of 1mM oxo- and sulfido-bridged Mo(V)–cysteine complexes at a Hg electrode in 0.1F $Na_2B_4O_7$. Scan rate = 0.1 V/sec. Inset, 2c: scan rate = 20 V/sec.

Table I. Voltammetric Data for Reduction of Oxo- and Sulfido-Bridged Mo(V) Complexes[a,b]

Compound	E_{pc} (V)	$E_{pc}-E_{p/2}$ (V)	ΔE_p (V)	$i_p/\nu^{1/2}C$ ($\mu A\ s^{1/2}/V^{1/2}_{mM}$)	k_1 (s^{-1})	
$Na_2Mo_2O_2(EDTA)$	-1.248	0.053	0.382	59.1	0	(stable product)
$Na_2Mo_2O_3S\ (EDTA)$	-1.079	0.052	0.172	60.1	0	(stable product)
$Na_2Mo_2O_2S_2(EDTA)$	-1.068	0.055	0.170	57.9	>0	(small)
$Na_2Mo_2O_4(Cys)_2$	-1.310	0.056	0.522	57.0	0.004	
$Na_2Mo_2O_3S(Cys)_2$	-1.213	0.079	0.296	49.2	0.9	
$Na_2Mo_2O_2S_2(Cys)_2$	-1.29(sh) (-1.39)c	(0.096)	(0.51)	(46.6)	100	

[a] From references (4), (5) and (6). [b] Data recorded at ν = 0.1 V/s in 0.1 F $Na_2B_4O_7$ at a hanging Hg drop electrode (A = 0.022 cm²); potentials in V vs. SCE. [c] Values in parentheses recorded at ν = 20 V/s.

slow dissociation reaction. Some dissociation also may occur
following reduction of $Mo_2O_3S(EDTA)^{2-}$. The dimeric Mo(III)-EDTA
products are clearly more stable than the analogous cysteine
compounds. We believe this greater stability is due primarily
to the fact that EDTA bridges both Mo centers and thereby increases the integrity of the binuclear unit. Another feature
observed during electrochemical experiments is that decomposition
of the initial Mo(III)$_2$-cysteine electrode products leads to
species which catalyze H^+ reduction at the mercury electrode.
This behavior is particularly pronounced for $Mo_2O_2S_2(Cys)_2^{2-}$, in
which case the voltammetric wave is observed as a shoulder on
the background discharge of H+. However, the characteristic
Mo(V)$_2$/Mo(III)$_2$ redox process for this compound is apparent at
faster scan rates (Inset, Figure 2c) where dissociation of the
Mo(III)$_2$ product is less extensive.

A mechanism for electrochemical reduction of the binuclear
Mo(V) complexes is shown in Figure 4. A detailed analysis of
buffer effects has shown that the initial step in the reaction is
a concerted 4-electron/4-proton transfer in which protonated
buffer species are involved in the transition state of the
electrode reaction (4,5). The protonated buffer species probably
interact with the terminal oxo groups of the $Mo_2O_4^{2+}$ unit and
facilitate coupled electron-proton transfer to produce coordinated aquo groups in the Mo(III) products. The buffer species
apparently replace water molecules within the Mo(III) coordination sphere soon after reduction of the Mo(V) dimer. For example,
the stable oxo-bridged Mo(III)$_2$-EDTA products display visible
absorption bands which shift with changes in buffer medium (5),
and a binuclear Mo(III) complex recently has been isolated in
which acetate, EDTA, and oxo groups simultaneously bridge the two
Mo atoms (8). Similar bridging by borate or phosphate oxyanions
(A) is represented in the electrode products in Figure 4.

Dissociation of the binuclear Mo(III) products to catalytically active species involves a complicated series of reactions.
The process appears to be intramolecular and to involve a buffer-
coordinated electrode product, since the rate of dissociation
depends on buffer type but not on pH, buffer concentration, or
addition of nucleophiles. For $Mo_2O_4(Cys)_2^{2-}$ it has been determined that the rate-determining step involves cleavage of one of
the μ-oxo bridge bonds to form a mono-oxo-bridged species. The
latter species is observed as a second anodic peak at slow voltammetric scan rates, but vanishes after total electrolysis. Further
steps in the sequence have not been fully delineated, but appear
at least to involve reaction to a further Mo(III) dimer (bridged
solely by oxo group or buffer anion) in equilibrium with Mo(III)
monomer (which is apparent from catalytic studies). The ultimate
electrode reaction products of $Mo_2O_4(Cys)_2^{2-}$ are non-electroactive
and have not been successfully characterized to date. However,
three bands (2 brown, 1 green) can be resolved by gel column

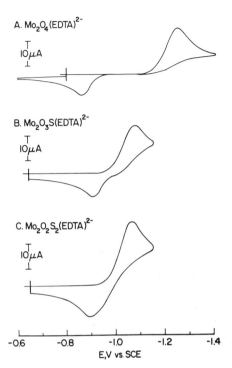

A. Mo2O4(EDTA)2-
10μA
B. Mo2O3S(EDTA)2-
10μA
C. Mo2O2S2(EDTA)2-
10μA
-0.6
-0.8
-1.0
-1.2
-1.4
E,V vs. SCE

Figure 3. Cyclic voltammetric curves for reduction of 1mM oxo- and sulfido-bridged Mo(V)–EDTA complexes at a Hg electrode in 0.1F $Na_2B_4O_7$. Scan rate = 0.1 V/sec.

Mo + 4HA + 4e⁻ ⇌ Mo Mo + 3A⁻
Keq
Mo Mo + 2H2O
k1
H2O
k2
further Mo(III)dimer
⇌ 2 Mo(III)-cys

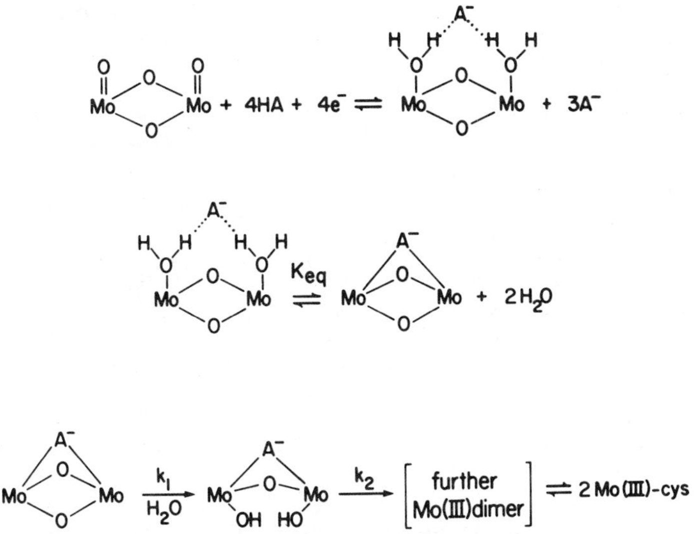

Figure 4. Proposed mechanism for electrochemical reduction of binuclear molybdenum(V) complexes

chromatography following exhaustive electrolysis of
$Mo_2O_4(Cys)_2^{2-}$. A mixture of products is consistent with the reaction scheme shown in Figure 4. The μ-oxo-μ-sulfido- and
di-μ-sulfido-bridged complexes appear to undergo similar dissociation reactions following reduction to the Mo(III) state.
Table I shows estimated values of the dissociation rate constant,
k_1, for all compounds in 0.1 $\underline{F}$ $Na_2B_4O_7$. These estimates were
made by cyclic voltammetry ($\underline{9}$) or double potential-step chrono-
coulometry ($\underline{10}$) assuming a pseudo first-order dissociation reaction (EC mechanism) with $k_{obs} = k_1K_{eq}/(1 + K_{eq}) \approx k_1$. More
detailed studies of the dissociation mechanism are currently
underway in our laboratory. As demonstrated in the following
section, this dissociation reaction plays a key role in the
generation of catalytically active species.

Electrocatalytic Reduction of Nitrogenase Substrates

Previous studies of the chemical model system ($\underline{1},\underline{3}$) have
failed to answer a number of important questions regarding the
nature of the active catalyst and the mechanism of catalytic
substrate reduction. In attempting to answer the latter two
points we have decided to investigate in detail the electrocatalytic reduction of a single substrate, acetylene, rather
than to survey the behavior of all known substrates of the
system. This choice is dictated largely by the fact that
acetylene provides more easily assayed products and is reduced
more rapidly in the model system than the true biological substrate, dinitrogen. While investigations with N_2 will provide
the ultimate biological relevance, our experiments with C_2H_2 have
provided significant information regarding the oxidation state
and properties of the reduced Mo catalysts.

Our experiments are carried out by controlled potential
coulometry at a stirred Hg pool cathode in a sealed cell equipped
with gas sampling ports ($\underline{7}$). Vapor phase samples are withdrawn
periodically for gas chromatographic analysis on Porapak N.
Two procedures are used to study the electrocatalytic reduction:
(A) a solution of $Mo_2O_4(Cys)_2^{2-}$ is reduced directly under 1 atm
of C_2H_2; (B) a solution of $Mo_2O_4(Cys)_2^{2-}$ is prereduced to the
Mo(III) state, the cell is purged with 1 atm C_2H_2, and potential
is reapplied. The latter procedure is more convenient because
rate plots are initially linear and intersect the origin. In
procedure A, 20-30 minutes pass before mass transport controlled
reduction of the complex is complete and acetylene is reduced at
a constant rate.

Typical behavior of the electrochemical part of the experiment is shown in Figure 5 using procedure B for the reduction of
C_2H_2 with $Mo_2O_4(Cys)_2^{2-}$ as catalyst. During controlled potential
reduction of the complex at -1.40 V, current falls to a relatively high steady-state value (5-10 mA) due to reduction of hydrogen
ion at the electrode surface catalyzed by the ultimate electrode

products of $Mo_2O_4(Cys)_2{}^{2-}$. The magnitude of the H^+ reduction current is proportional to the square-root of initial dimer concentration, thus indicating catalysis by a monomeric species. After the reduced complex is purged with acetylene and potential is reapplied, the current rises to a much higher steady-state value than during the previous electrolysis. The magnitude of this current also increases with negative potential following an initial reduction at -1.40 V.

Application of potential to the reduced molybdenum solution under 1 atm C_2H_2 results in the reduction of acetylene by an apparent first-order process (i.e., $log[C_2H_2] = -kt + const.$) and linear production of ethylene and ethane over a 2-3 hour period. Distribution of these species during a typical experiment is shown in Figure 6. The $C_2H_4:C_2H_6$ product ratio is about 4 or 5:1 and varies only slightly with changes in experimental conditions. A considerable quantity of H_2 gas also is evolved concurrently with acetylene reduction. Closer inspection of Figure 6 reveals that amounts of C_2H_4 and C_2H_6 produced do not equal the quantity of C_2H_2 reduced. An additional hydrocarbon product is 1,3-butadiene (C_4H_6), which is produced at about 40% the level of ethylene (i.e., $C_2H_4:C_4H_6:C_2H_6{\approx}5:2:1$). Butadiene also has been discovered to be the major product of acetylene reduction in the $Mo_2O_4(Cys)_2{}^{2-}/C_2H_2/BH_4{}^-$ chemical model system (11), but it is not produced upon reduction of C_2H_2 by the enzyme. With inclusion of C_4H_6 a suitable hydrocarbon balance is obtained for the chemical model system (11), but the product balance for our electrocatalytic system falls as much as 60% short of the quantity of acetylene reduced. Thus, additional and as yet undetected products must be produced in the electrochemical system. We did not discover the presence of butadiene or the discrepancy in hydrocarbon balance early enough to take these factors into account in all aspects of our investigation. Consequently, most results for the electrocatalytic system are based either on the rate of C_2H_2 reduction or the rate of C_2H_4 and C_2H_6 production.

Our present results provide strong evidence that a monomeric Mo(III) complex is the catalytically active species in chemical models for nitrogenase based on molybdenum-sulfhydryl complexes. The electrode reaction mechanism studies of $Mo_2O_4(Cys)_2{}^{2-}$ and the analogous oxo- and sulfido-bridged complexes with cysteine and EDTA establish that these compounds are reduced directly to Mo(III) products with no evidence of the intermediate Mo(IV) oxidation state. Involvement of Mo(III) is established by the fact that C_2H_2 reduction occurs subsequent to the reduction of these complexes by two electrons per molybdenum. In addition, application of negative potential to solutions containing equimolar Mo(III) (added as K_3MoCl_6) and cysteine catalyzes acetylene reduction at approximately the same rate (see Table II, procedure C). It is not likely that reduction proceeds beyond Mo(III) in formation of the active catalyst, because electrochemically reduced solutions of $Mo_2O_4(Cys)_2{}^{2-}$ or solutions of Mo(III) plus

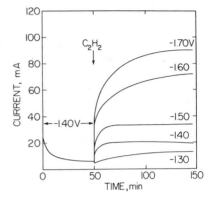

Figure 5. Current-time curves observed during electrocatalytic reduction of acetylene using procedure B. Experimental conditions: 1.70mM Na$_2$Mo$_2$O$_4$(Cys)$_2$, 0.1F Na$_2$B$_4$O$_7$, cell purged with 1 atm C$_2$H$_2$ and potential reapplied as indicated.

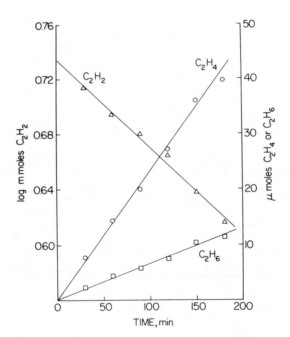

Figure 6. Product-time and reactant-time behavior during electrocatalytic reduction of acetylene using procedure B. Experimental conditions: 1.70mM Na$_2$Mo$_2$O$_4$(Cys)$_2$, 0.1F Na$_2$B$_4$O$_7$, E$_{app}$ = −1.40 V, 1 atm C$_2$H$_2$.

Table II. Electrocatalytic Reduction of Acetylene with
$Na_2Mo_2O_4(Cys)_2$

Proce-dure [a]	E_{app} (V)	Conc. (mM)	pH [b]	Rate (μmol C_2H_4/ min)	C_2H_4: C_2H_6
A	-1.40	1.00	9.6	0.204	4.8
A	-1.40	1.00	9.2	0.264	4.9
A	-1.40	1.00	8.8	0.343	4.5
B	-1.40	0.30	9.2	0.100	5.0
B	-1.40	1.00	9.2	0.195	4.5
B	-1.40	1.70	9.2	0.240	4.9
B	-1.40	3.00	9.2	0.313	4.7
B	-1.30	1.70	9.2	0.136	3.9
B	-1.40	1.70	9.2	0.240	4.9
B	-1.50	1.70	9.2	0.353	3.9
B	-1.60	1.70	9.2	0.498	4.2
B	-1.70	1.70	9.2	0.708	4.2
B	-1.35	1.00	8.3(P)	0.139	2.4
B	-1.27	1.00	9.2(A)	0.106	2.9
C	-1.40	1.00	9.2	0.133	4.7
D	no E	1.00	9.2	0.003	-

[a]Procedures: A. Complex reduced under 1 atm C_2H_2
　　　　　　　B. Complex prereduced at -1.40 V, cell purged with
　　　　　　　　 1 atm C_2H_2, potential reapplied as indicated
　　　　　　　C. Potential applied to 2mM K_3MoCl_6 + 2mM cysteine
　　　　　　　　 under 1 atm C_2H_2
　　　　　　　D. Complex prereduced at -1.40 V, purged with 1 atm
　　　　　　　　 C_2H_2, and allowed to stand without potential
　　　　　　　　 applied

[b]
All solutions contain 0.1 $\underline{F}$ $Na_2B_4O_7$ except P = phosphate (0.5 $\underline{F}$)
and A = ammonia (0.25 $\underline{F}$)

cysteine exhibit no voltammetric reduction peaks.

A number of control experiments have been performed and demonstrate that Mo(III) plus an additional source of electrons (or H_2) is required for C_2H_2 reduction. For example, electrolysis of solely C_2H_2 at 1 atm in borate buffer at very negative potentials, even in the presence of H_2-evolving catalysts such as cysteine, produces no reduced hydrocarbons. If the $Mo_2O_4(Cys)_2{}^{2-}$ complex is reduced electrochemically, purged with 1 atm C_2H_2, and allowed to stand, only minimal C_2H_4 is produced (Table II, procedure D). When potential is reapplied, catalysis resumes at the normal rate.

A sampling of data is shown in Table II illustrating the effects of various experimental conditions and components on the electrocatalytic reduction of acetylene. Notable factors which increase the rate of C_2H_4 production are decreasing pH, increasing concentration of complex, and increasing negative potential. Several experiments run in phosphate and ammonia buffer indicate that the rate of C_2H_4 production is decreased slightly relative to borate buffer.

The effect of $Mo_2O_4(Cys)_2{}^{2-}$ concentration on the rates of C_2H_4 and C_2H_6 production is shown in Figure 7. These square-root dependences on initial concentration of complex demonstrate that the active catalyst is a monomeric species in equilibrium with a larger fraction of dimeric material. The same result is found in the chemical model system regarding C_2H_4 production (3,12), but a linear dependence of C_2H_6 formation on $[Mo_2O_4(Cys)_2{}^{2-}]^{1/2}$ has not been noted previously. Production of a monomeric catalyst could be achieved through a sequence of chemical steps as outlined in Figure 4. Although we have not confirmed this mechanism in the detail depicted, sequential dissociation of the initial Mo(III)$_2$ electrode product through one or more dimeric intermediates is consistent with the electrochemical results, observation of several species in product isolation attempts, and observation of an equilibrium concentration of monomeric catalyst.

Dissociation of Mo(III) atoms following reduction of the binuclear center is clearly an important step in the generation of an active catalyst. For this reason the entire series of oxo- and sulfido-bridged Mo(V)-cysteine and EDTA complexes described earlier has been examined in the acetylene reduction experiment. Results are shown in Table III, and indicate that sulfur bridging atoms and ligand play an important role in producing an active catalyst and in the rate and mechanism of C_2H_2 reduction. The sulfido-bridged cysteine complexes, $Mo_2O_3S(Cys)_2{}^{2-}$ and $Mo_2O_2S_2(Cys)_2{}^{2-}$, are reduced to Mo(III)$_2$ products which dissociate rapidly on the voltammetric time scale. These products, however, provide only a marginal increase in the rate of C_2H_2 reduction and similar C_2H_4:C_2H_6 ratios of 4 or 5:1. The di-μ-oxo EDTA complex, $Mo_2O_4(EDTA)^{2-}$, is reduced to a stable binuclear product which is completely ineffective in acetylene reduction.

Table III. Electrocatalytic Reduction of Acetylene with Various
Oxo- and Sulfido-Bridged Molybdenum(V) Complexes[a]

Compound	Rate (μmol C_2H_4 + C_2H_6/min)	C_2H_4:C_2H_6
$Na_2Mo_2O_4(Cys)_2$	0.302	3.9
$Na_2Mo_2O_3S(Cys)_2$	0.323	4.0
$Na_2Mo_2O_2S_2(Cys)_2$	0.324	4.8
$Na_2Mo_2O_4(EDTA)$	0.0	-
$Na_2Mo_2O_3S(EDTA)$	0.048	0.20
$Na_2Mo_2O_2S_2(EDTA)$	0.330	0.33

a) All solutions contain 1.70 mM complex in 0.1 $\underline{F}$ $Na_2B_4O_7$, E_{app} =
-1.40 V, 1 atm C_2H_2, procedure B (see footnote a, Table II).

$Mo_2O_3S(EDTA)^{2-}$, which shows some activity towards C_2H_2 reduction,
may dissociate slightly after electrochemical reduction. The
di-μ-sulfido complex, $Mo_2O_2S_2(EDTA)^{2-}$, however, dissociates slowly
but quite evidently following its electrochemical reduction.
This reduced material catalyzes production of C_2H_4 and C_2H_6 at a
rate equal to the corresponding cysteine compound and yields an
inverted product ratio of C_2H_6:C_2H_4 = 3:1.

Figure 8 shows the significant effect of electrode potential
on the electrocatalytic process. The rates of C_2H_4 and C_2H_6 production and acetylene reduction increase exponentially with potential, as does the steady-state current following application of
potential to an acetylene-purged solution. One experimental observation which parallels this behavior is the catalytic evolution of H_2 at the mercury electrode, which occurs concurrently
with acetylene reduction and also in the presence of the Mo(III)
electrode products alone. During a typical experiment hydrogen
evolution may account for 30-40% of the total coulombs passed,
the remainder going to C_2H_2 reduction. We believe it is premature
to dismiss hydrogen evolution as an experimental artifact. Evolution of H_2 in the absence of substrates other than H^+ is an important feature of the chemical model system (3,12) and of nitrogenase enzyme (13). In the chemical model, for example: (a)
$NaBH_4$ is a more effective reductant than $Na_2S_2O_4$ and also catalyzes H_2 evolution more strongly (3), and (b) addition of cocatal-
ysts such as methyl viologen (14) and $Fe_4S_4(SR)_4^{2-}$ clusters

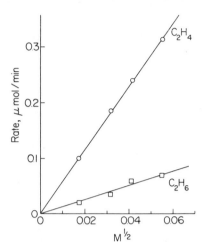

Figure 7.   Rates of ethylene and ethane production as a function of square root of $Na_2Mo_2O_4(Cys)_2$ concentration. Other experimental conditions as in Figure 6.

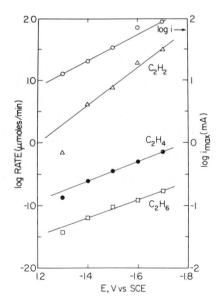

Figure 8.   Effect of electrode potential on catalytic current, acetylene reduction rate, and ethylene and ethane production rates during electrocatalytic reduction of C_2H_2 using procedure B. Experimental conditions as in Figure 5.

($\underline{12}$,$\underline{15}$), which increase the C_2H_2 reduction rate, also stimulate the production of H_2. The effect of electrode potential indicates that an important relationship exists between H_2 evolution and C_2H_2 reduction in the electrocatalytic system. Clearly, an electrode activated rather than a bulk solution process is involved in the rate-determining step of acetylene reduction.

Four possible mechanisms which have been considered for the electrochemically catalyzed reduction of acetylene are listed in Table IV. In each case formation of a Mo(III)-C_2H_2 adduct is assumed to be the initial step in the mechanism, because the requirement of molybdenum indicates that some interaction between the reduced Mo species and acetylene must occur during catalysis. It is not possible to confirm or reject any of these mechanisms with certainty. Mechanism D, however, is contradicted least by present evidence. This mechanism is viewed as an electrocatalytic hydrogenation in which the principal function of the Mo(III) catalyst is to bind acetylene for insertion of hydrogen atoms produced at the electrode surface. Hydrogen atoms are the product of hydrogen ion reduction catalyzed by the Mo(III)-cysteine

Table IV. Possible Mechanisms for Electrocatalytic Reduction of Acetylene

---

$$Mo(III) + C_2H_2 \rightleftarrows Mo(III)\text{-}C_2H_2 \quad \text{followed by:}$$

A. Homogeneous Redox Reaction

$$Mo(III)\text{-}C_2H_2 + 2H^+ \rightarrow Mo(V) + C_2H_4$$

$$Mo(V) + 2e^- \rightleftarrows Mo(III)$$

B. Electrochemical Reduction of Adduct

$$Mo(III)\text{-}C_2H_2 + 2H^+ + 2e^- \rightarrow Mo(III) + C_2H_4$$

C. Homogeneous Hydrogenation

$$Mo(III)\text{-}C_2H_2 + H_2 \rightarrow H\text{-}Mo(III)\text{-}C_2H_2 + H^+$$

$$H\text{-}Mo(III)\text{-}C_2H_2 + H^+ \rightarrow Mo(III) + C_2H_4$$

D. Electrocatalytic Hydrogenation

$$\text{or} \begin{cases} Mo(III)\text{-}Cys + 2H^+ + 2e^- \rightarrow Mo(III)\text{-}Cys + 2H\cdot \\ Mo(III)\text{-}C_2H_2 + 2H^+ + 2e^- \rightarrow Mo(III)\text{-}C_2H_2 + 2H\cdot \end{cases}$$

$$Mo(III)\text{-}C_2H_2 + 2H\cdot \rightarrow Mo(III) + C_2H_4$$

complex with or without C_2H_2 bound to it. This process is much
the same as the traditional evolution of H_2 at mercury electrodes
catalyzed by transition metal complexes with sulfur-containing
ligands (16), and would exhibit an exponential dependence on
potential. Catalysis presumably occurs when a Mo(III)-coordi-
nated sulfur atom (from either cysteine ligand or bridging S
group) is protonated to form an -SH$^+$ species and then reduced:

$$Mo(III)-S: + H^+ \rightleftarrows Mo(III)-SH^+ \qquad (6)$$

$$Mo(III)-SH^+ + e^- \rightarrow Mo(III)-SH \cdot \qquad (7)$$

These species could react to form H_2

$$2\ Mo(III)-SH \cdot \rightarrow 2\ Mo(III)-S: + H_2 \qquad (8)$$

or, in the presence of Mo(III)-C_2H_2 adduct, react to form C_2H_4:

$$2\ Mo(III)-SH \cdot + Mo(III)-C_2H_2 \rightarrow 2\ Mo(III)-S: + Mo(III) + C_2H_4 \qquad (9)$$

The alternative mechanisms in Table IV are contradicted by
at least one piece of experimental evidence. In mechanisms A
and B the regeneration of catalyst and reduction of adduct are
accomplished through diffusion-limited electrochemical processes.
It is difficult to imagine how these electrochemical steps could
be so irreversible as to display the observed potential dependence over a range of 400-500 mV. Mechanism A also is discounted
by the fact that addition of C_2H_2 to solutions of Mo(III) does
not lead to substantial formation of C_2H_4. Mechanism C is
similar to known homogeneous hydrogenation reactions catalyzed by
transition metal complexes (17). A similar mechanism involving
a hydridic intermediate has been suggested for the chemical
system (18). This mechanism does not seem attractive in the
electrocatalytic system, because, as the quantity of H_2 in the
cell is constantly increasing during electrolysis, the rate of
catalytic acetylene reduction remains constant.

Ethane and 1,3-butadiene are two additional products of electrocatalytic acetylene reduction. Ethane is produced in constant
proportion to ethylene under a variety of experimental conditions,
and ethylene itself is not reduced in the catalytic system.
Therefore, a separate binding reaction between Mo(III) and
ethylene does not take place before reduction to C_2H_6. Also, it
is not likely that a dimeric Mo(III) species is responsible for
reduction of C_2H_2 to C_2H_6 because ethane formation is linearly
dependent on $[Mo_2O_4(Cys)_2^{2-}]^{1/2}$ (Figure 7) and the Mo(III) dimer
produced by reduction of $Mo_2O_4(EDTA)^{2-}$ is completely inactive in
acetylene reduction (Table III). We believe it is more likely
that about 20-25% of the time two additional hydrogen atoms are
inserted into a bound acetylene molecule before dissociation from
the Mo(III) catalyst takes place:

$$Mo(III)-C_2H_2 + 2H\cdot \rightarrow Mo(III)-C_2H_4 \tag{10}$$

$$Mo(III)-C_2H_4 + 2H\cdot \rightarrow Mo(III) + C_2H_6 \tag{11}$$

Formation of butadiene can occur if C_2H_2 interacts with the Mo(III)-C_2H_2 adduct before reduction.

$$Mo(III) \underset{CH}{\overset{CH}{\left\langle \;\right.}} \underset{\cdots CH}{\overset{\|}{}} + \underset{CH}{\overset{CH}{\|\|}} \xrightarrow{2H^+, 2e^-} Mo(III) + H_2C=CH-CH=CH_2 \tag{12}$$

Subsequent polymerization reactions or electrochemical reduction of butadiene could explain the inability to obtain a total hydrocarbon balance in the electrochemical system. Reaction 12 could well be favored at the high C_2H_2 partial pressures used in this work. The effect of acetylene partial pressure on product distribution has not been studied in either the chemical or electrochemical catalytic systems.

Discussion

Several years' study of these model systems has impressed upon us the complicated nature of molybdenum solution chemistry. Specific and unique effects of bridging atom, ligand, solution environment, and buffer salt are evident in the chemical, electrochemical, and catalytic properties of these compounds. Such complexity, incompletely understood, makes the extrapolation of results to an equally complicated biological system somewhat tenuous. However, we believe our electrochemical studies have provided results which are directly useful in understanding the behavior of chemical models for nitrogenase based on molybdenum-sulfhydryl complexes. These results also contribute to the general knowledge of molybdenum chemistry which is necessary in interpreting the behavior of the enzyme. A number of significant points emanating from these electrochemical studies are discussed below.

1. **Molybdenum Oxidation State.** All binuclear Mo(V) compounds we have examined undergo $Mo(V)_2 \rightarrow Mo(III)_2$ reduction under aqueous solution conditions comparable to those used in studies of the chemical model and the enzyme itself. There is no evidence of the intermediate Mo(IV) oxidation state. It seems likely, therefore, that Mo(III) and not Mo(IV) is the oxidation state of the active catalyst in the Mo-cysteine model system. Despite recent comments to the contrary (19), we also believe that Mo(III) is a strong candidate for the oxidation state of the reduced Mo center in nitrogenase. Other evidence also supports the possible importance of Mo(III) in the molybdenum-containing reductases: the reduction of N_2 and C_2H_2 catalyzed by inorganic Mo(III) species over a range of temperatures and pressures (20,21), construction

of a successful model for nitrate reduction based on the hexa-
aquomolybdenum(III) cation, $Mo(H_2O_6)^{3+}$ (22), and tentative observation of Mo(III) epr signals in nitrate reductase (23-25).
Evidence for the molybdenum oxidation state in reduced nitrogenase may be provided by x-ray absorption edge spectroscopy (26),
but definitive results are not yet available.

   2. Coupled Electron-Proton Transfer. Reduction of the binuclear complexes proceeds by coupled transfer of four electrons
and four protons in a single step. This observation is significant in view of a recent proposal (27) that coupled electron-
proton transfer in multiples of two is an important feature of
molybdenum-containing enzymes. Furthermore, the unique buffer
effects observed in the electrode reaction mechanism suggest
that oxomolybdenum species may facilitate proton transfer to or
from substrates in a highly specific manner.

   3. Dissociation of the Binuclear Center. Dissociation of
Mo(III) atoms following reduction of the binuclear Mo(V) unit
appears to be an essential step in the generation of an active
catalyst. Only those compounds which show evidence of dissociation following electrochemical reduction are effective in the
catalytic reduction of C_2H_2 and H^+. On the other hand, the
dimeric Mo(III) reduction product of $Mo_2O_4(EDTA)^{2-}$ does not
catalyze reduction of C_2H_2, even though it is an extremely strong
reducing agent ($E^{0'}$ = -1.06 V vs. SCE in 0.1 F $Na_2B_4O_7$). Also,
it is apparent that dissociation of the binuclear center occurs
after, not prior to (3), reduction to the Mo(III) state, and
that sulfur bridging atoms and ligands which do not bridge the
two Mo centers increase this tendency for dissociation.

   4. Sulfur Bridging Atoms. Sulfur bridging atoms increase
the reversibility of electron transfer in the binuclear Mo(V)
center and the ease of dissociation of binuclear Mo(III) units.
Both features enhance the catalytic properties of the compounds
we have studied. Thus, sulfur bridging may be an important and
desirable feature of Mo enzyme model chemistry. Several instances have been noted wherein sulfido bridging has imparted
unusual stability to binuclear Mo(V) species (28,29), and it has
been suggested that such a feature may be undesirable in terms of
Mo enzyme model chemistry (29). In the Mo(III) oxidation state,
however, sulfido bridging enhances the reactivity of the binuclear unit and thus improves catalytic activity. Presence of
sulfur in the Mo(III) coordination sphere may increase the labil-
ity of this oxidation state and permit more facile binding of substrates. Similar increases in ease of substitution promoted by
thiol ligands have been noted recently in chromium(III)
chemistry (30-32).

5. Biological Function of Molybdenum. The primary function
of the reduced molybdenum species may be simply to bind rather
than transfer electrons to the substrate. In our electrochemical
studies a number of mechanisms appear to be at least as probable
as one involving a bulk solution Mo(V)/Mo(III) redox cycle. In
the chemical model systems these alternative mechanisms could be
translated to ones in which the chemical reductant [NaBH$_4$,
Na$_2$S$_2$O$_4$ or Fe$_4$S$_4$(SR)$_4^{4-}$] transfers electrons (or H$_2$) to
molybdenum-bound C$_2$H$_2$ without need of reoxidizing the Mo(III)
center. In nitrogenase, a similar process can be pictured in
which the molybdenum center is first reduced to its substrate-
binding oxidation state. Once bound, the substrate is reduced
by a flow of electrons or reactive hydrogen from a proximal site
(presumably Fe$_4$S$_4$-type ferredoxin) and then released, leaving the
molybdenum site in its reduced state. Transfer of two electrons
and two protons or transfer of two hydrogen atoms would be virtually equivalent mechanisms if the two sites were close together
or bridged by an atom such as sulfur which could facilitate both
proton and electron transfer.

Acknowledgment

This research has been supported by the National Science
Foundation under Grant GP-38442X. We are particularly grateful
to Drs. W. E. Newton, E. I. Stiefel, J. W. McDonald and J. L.
Corbin of the Charles F. Kettering Research Laboratory, Yellow
Springs, Ohio for many fruitful discussions and for disclosing
to us their discovery of butadiene product in the chemical model
system prior to publication.

Literature Cited

1. Schrauzer, G.N., Angew. Chem. Int. Ed., (1975) 14, 514, and
references therein.
2. Kay, A. and Mitchell, P. C. H., J. Chem. Soc. A, (1970),
2421.
3. Schrauzer, G. N. and Doemeny, P. A., J. Amer. Chem. Soc.,
(1971) 93, 1608.
4. Ott, V. R. and Schultz, F. A., J. Electroanal. Chem., (1975)
61, 81.
5. Ott, V. R. and Schultz, F. A., J. Electroanal. Chem., (1975)
59, 47.
6. Ott, V. R., Swieter, D. S. and Schultz, F. A., manuscript
in preparation.
7. Ledwith, D. A. and Schultz, F. A., J. Amer. Chem. Soc.,
(1975) 97, 6591.
8. Kneale, G. G., Geddes, A. J., Sasaki, Y., Shibahara, T. and
Sykes, A. G., J. Chem. Soc. Chem. Commun., (1975), 356.
9. Nicholson, R. S. and Shain, I., Anal. Chem., (1964) 36, 706.

10. Ridgway, T. H., Van Duyne, R. P. and Reilley, C. N., J. Electroanal. Chem., (1972) 34, 267, 283.
11. Corbin, J. L., Pariyadath, N. and Stiefel, E. I., J. Amer. Chem. Soc., in press.
12. Tano, K. and Schrauzer, G. N., J. Amer. Chem. Soc., (1975) 97, 5404.
13. Hardy, R. W. F., Burns, R. C. and Parshall, G. W., Advan. Chem. Series, (1971) 100, 219.
14. Ichikawa, M. and Meshitsuka, S., J. Amer. Chem. Soc., (1973) 95, 3411.
15. Schrauzer, G. N., Kiefer, G. W., Tano, K. and Doemeny, P. A., J. Amer. Chem. Soc., (1974) 96, 641.
16. Mairanovskii, S. G., "Kinetic and Catalytic Waves in Polarography," Plenum Press, New York, 1968.
17. James, B. R., "Homogeneous Hydrogenation," Wiley, New York, 1973.
18. Khrushch, A. P., Shilov, A. E. and Vorontsova, T. A., J. Amer. Chem. Soc., (1974) 96, 4987.
19. Wentworth, R. A. D., Coordin. Chem. Rev., (1976) 18, 1.
20. Denisov, N. T., Shuvalov, V. F., Shuvalova, N. I., Shilova, A. K. and Shilov, A. E., Dokl. Akad. Nauk SSSR, (1970) 195, 879.
21. Shilov, A. E., Denisov, N. T., Efimov, O. N., Shuvalov, V. F., Shuvalova, N. D. and Shilova, A. K., Nature, (1971) 231, 460.
22. Ketchum, P. A., Taylor, R. C. and Young, D. C., Nature (1976), 259, 202.
23. Forget, P. and DerVartanian, D. V., Biochim. Biophys. Acta, (1972) 256, 600.
24. DerVartanian, D. V. and Forget, P., Biochim. Biophys. Acta, (1975) 379, 74.
25. Orme-Johnson, W. H., Jacob, G., Henzl, M. and Averill, B. A., ACS Centennial Meeting, New York, 1976, Abstr. INOR-137.
26. Cramer, S. P., Eccles, T. K., Kutzler, F. W., Hodgson, K. O. and Mortenson, L. E., J. Amer. Chem. Soc., (1976) 98, 1287.
27. Stiefel, E. I., Proc. Nat. Acad. Sci. U.S.A., (1973) 70, 988.
28. Spivack, B. and Dori, Z., J. Chem. Soc. Chem. Commun., (1973), 909.
29. Newton, W. E., Corbin, J. L., Bravard, D. C., Searles, J. E. and McDonald, J. W., Inorg. Chem., (1974) 13, 1100.
30. Weschler, C. J. and Deutsch, E., Inorg. Chem., (1973) 12, 2682.
31. Ramasami, T. and Sykes, A. G., Inorg. Chem., (1976) 15, 1010.
32. Asher, L. E. and Deutsch, E., Inorg. Chem., (1976) 15, 1531.

Manganese(II) and -(III) 8-Quinolinol Complexes. Redox Model for Mitochondrial Superoxide Dismutase

JOHN K. HOWIE, MARK M. MORRISON, and DONALD T. SAWYER

Department of Chemistry, University of California, Riverside, Calif. 92502

The discovery in 1969 ($\underline{1}$) that superoxide ion, O_2^-, is a common respiratory intermediate of aerobic organisms with its concentration controlled by superoxide dismutase (SOD) has revolutionized the interpretation of biological oxidation-reduction processes. A subsequent discovery was a manganese-containing version of superoxide dismutase which can be isolated from bacterial sources ($\underline{2},\underline{3}$) and from mitochondria ($\underline{4}$) as well as the originally discovered copper-zinc form from erythrocytes. Little is known about manganese SOD. The crystal structure has not been determined and there is still controversy as to whether the manganese SOD contains one or two manganese atoms per enzyme molecule ($\underline{2},\underline{4}$). The exact role of the manganese atom(s) in the enzyme, the oxidation state(s) of the manganese atom(s), the degree of association of the two manganese atoms, if two are indeed present, the type of ligands coordinated to the manganese atom(s), and the stereochemistry around the metal(s) are not known.

Although little is known about the structure and properties of manganese SOD, its catalytic reactions with superoxide ion can be represented by ($\underline{5}$)

$$E + O_2^- \longrightarrow E^- + O_2 \qquad (1)$$

$$E^- + O_2^- \xrightarrow{\;H^+\;} E + H_2O_2 \qquad (2)$$

$$E^- + O_2^- \longrightarrow E^{2-} + O_2 \qquad (3)$$

$$E^{2-} + O_2^- \xrightarrow{\;H^+\;} E^- + H_2O_2 \qquad (4)$$

Such a mechanism requires that the manganese exist in three different oxidation states if the enzyme contains only one metal atom per molecule. However,

only two formal oxidation states are required if the
enzyme contains two metal atoms. Because the manganese-containing enzymes catalyze oxidation-reduction
reactions, electrochemical methods are particularly
attractive for the study of the redox behavior of
manganese complexes, both alone and in the presence
of substrate. The further objective is to characterize the structures and coordination chemistry of
these complexes in solution by use of spectroscopic
as well as other physical characterization techniques.

The goal of the present research is the identification and characterization of manganese complexes
that mimic the enzyme in reactions represented by
Equations 1-4 and that can serve as models for manganese SOD. This paper reports on studies of mangan-
ese(II) and -(III) 8-quinolinol complexes acting as
model compounds.

Experimental

Measurements and Materials. Cyclic voltammetric
experiments were performed using a versatile instrument constructed from Philbrick solid-state operational amplifiers (6). The controlled potential
electrolysis experiments were performed using a
Wenking Model 61RH potentiostat and integrating the
current vs. time curve using a K&E Model 62005 com-
pensating polar planimeter. The electrochemical cell
employed in all electrochemical experiments was
described previously (7). A Beckman Model 39273
platinum inlay electrode was used as the working
electrode for cyclic voltammetry and a platinum gauze
electrode was employed as the working electrode in
the coulometric experiments. The reference electrode
was composed of a Ag/AgCl electrode in aqueous tetra-
methylammonium chloride solution (0.000 V vs. SCE)
and a glass bridge tube which made contact with the
bulk solution through a cracked glass-bead junction.
The platinum flag auxiliary electrode was isolated
from the bulk solution by a fine porosity frit.

Dimethyl sulfoxide (DMSO) (J. T. Baker analyzed
reagent grade) had a water content of 0.02 to 0.06%
as specified by the manufacturer. Pyridine (Burdick
and Jackson) contained 0.009% water and acetonitrile
(MC/B Spectroquality grade) contained a maximum of
0.02% water. The solvents were degassed with argon
in the electrochemical cell prior to the addition of
the compound to be studied. Tetraethylammonium
perchlorate (TEAP) was used as the supporting electro-

lyte in a 50-to 100-fold excess over the concentration
of the electroactive species.

Solid-state magnetic susceptibility values
were determined by the Guoy method and solution magnetic susceptibility values were determined by
the nuclear magnetic resonance (nmr) method (8).
Conductivity measurements were made with an Ind. Instr.
Model RC16B conductivity bridge and a dip cell. Molecular weight determinations were made cryoscopically
in DMSO solution using a Hewlett-Packard Model 2801A
Quartz thermometer to measure temperature changes.
The apparatus was calibrated with benzil and all solutions were prepared from the same freshly opened
bottle of DMSO. The solutions were protected from
contamination by water in the air by an atmosphere of
dry argon.

8-Quinolinol was obtained from MC/B and mangan-
ous acetate tetrahydrate, $Mn^{II}(OAc)_2 \cdot 4H_2O$, was obtained from Alfa. Methanol was reagent grade and
used without further purification. Argon was dried
by passing it through a column packed with Aquasorb
(Mallinckrodt). 1.00 $\underline{M}$ $HClO_4$ in water and 1.42 $\underline{M}$
tetraethylammonium hydroxide in methanol (Eastman)
were used in the electrochemical experiments where
hydrogen and hydroxide ions were employed.

Preparation of the Complexes

1. Bis(8-quinolinolato)manganese(II) dihydrate,
    $Mn^{II}Q_2 \cdot 2H_2O$. The compound was prepared by the
    reaction between 10 g (0.041 mole) $Mn^{II}(OAc)_2 \cdot$
    $4H_2O$ and 11.8 g (0.082 mole) HQ in 250 ml of
    deaerated 1:1 methanol/water. The yellow product
    was filtered under argon, washed with deaerated
    water and methanol, and dried <u>in vacuo</u> at room
    temperature for 2 hr. Elemental analysis: Calcd.
    for $MnC_{18}N_2H_{16}O_4$: Mn, 14.49; C, 57.00; N, 7.39;
    and H, 4.26. Found: Mn, 14.27; C, 56.63; N, 7.04;
    and H, 4.31.

2. μ-Oxo-bis(8-quinolinolato-8-quinolinol)manganese-
    (III) dimethanol, $Mn_2^{III}OQ_4(HQ)_2 \cdot 2CH_3OH$. The
    compound was prepared by the reaction of air with
    a saturated solution of $Mn^{II}Q_2 \cdot 2H_2O$ in 1:1 meth-
    anol/water. The product which formed as black
    crystals was filtered, washed with water and
    methanol, and dried <u>in vacuo</u> at room temperature
    for 2 hr. Elemental analysis: Calcd. for
    $Mn_2C_{56}N_6H_{46}O_9$: Mn, 10.39; C, 63.64; N, 7.95; and
    H, 4.38. Found: Mn, 10.14; C, 63.70; N, 7.76;

and H, 4.30.

3. Bis(8-quinolinolato)magnesium(II), $Mg^{II}Q_2$, and bis(8-quinolinolato)zinc(II), $Zn^{II}Q_2$. Both of the compounds were prepared by the procedure used to synthesize $Mn^{II}Q_2 \cdot 2H_2O$.

4. Superoxide ion, O_2^-. Superoxide ion was generated in situ in DMSO and pyridine solutions by controlled potential electrolysis at -1.00 V at a gold foil electrode of oxygen-saturated solutions. The solutions were degassed with argon, prereduced at -1.00 V for 10 min, and then saturated with oxygen. The oxygen flow was continued throughout the electrolysis. To avoid the formation of protons during electrolysis which may migrate into the working electrode compartment, the auxiliary electrode compartment was filled with a 40% solution of tetraethylammonium hydroxide in water. The solutions were degassed with argon prior to use.

Results

Dissociation and Magnetic Susceptibilities of Manganese Complexes. Conductance measurements indicate that $Mn^{II}Q_2 \cdot 2H_2O$ in DMSO solution is about 10% dissociated into a 1-to-1 electrolyte and that solutions of $Mn_2^{III}OQ_4(HQ)_2 \cdot 2CH_3OH$ are not dissociated into ionically-conducting species. The molecular weight determinations indicate that $MnQ_2 \cdot 2H_2O$ is about 80% dissociated into MnQ_2 and H_2O.

The manganese(III) 8-quinolinol complex contains a high spin d^4 manganese ion in the solid state ($\mu = 5.0 \pm 0.1$ B.M.), but in DMSO solution the complex exhibits a decreased magnetic moment ($\mu_{corr} = 4.56 \pm 0.10$ B.M.) The magnetic moment is close to the spin only value in acetonitrile solution($\mu_{corr} = 4.8 \pm 0.2$ B.M.) and in pyridine solution ($\mu_{corr} = 4.91 \pm 0.07$ B.M.).

Electrochemistry of Mn(II)- and Mn(III)-8-Quinolinol Complexes. Cyclic voltammograms of $Mn^{II}Q_2 \cdot 2H_2O$ and $Mn_2^{III}OQ_4(HQ)_2$ in DMSO solution appear in Figure 1. The two compounds share common redox products. $Mn^{II}Q_2 \cdot 2H_2O$ is not reduced but is oxidized at +0.16 and +0.75 V. The +0.16 V peak is a reversible one-electron per manganese oxidation based on peak currents; however, controlled potential coulometry at +0.25 V reveals that on the longer coulometric time scale (~20 min) the oxidation is only a 0.5 electron process. If an equivalent of hydroxide ions is added at this point and the electrolysis is continued, another 0.25 electron per manganese atom is transferred.

Reversing the cyclic voltammetric scan direction after the oxidation at +0.16 V reveals another cathodic peak at -0.31 V. This peak is observed also with $Mn_2^{III}OQ_4(HQ)_2$. For both complexes, controlled potential coulometry at -0.50 V indicates that this is a one-electron (per manganese atom) reduction. For $Mn_2^{III}OQ_4(HQ)_2$, as seen by cyclic voltammetry, most of the reduction product formed at -0.50 V is reoxidized at -0.16 V with a small additional amount being reoxidized at +0.16 V. For $Mn^{II}Q_2 \cdot 2H_2O$, previously electrolyzed at +0.25 V, the situation is reversed with respect to the product yields at -0.16 and +0.16 V. Addition of 1 equivalent of protons (per mole of manganese atoms) to the original $Mn_2OQ_4(HQ)_2$ solutions after electrolysis at -0.5 V yields cyclic voltammograms identical to those observed after the electrolysis sequence described above for solutions of $Mn^{II}Q_2$.

$Mn^{II}Q_2$ has a second large anodic peak at +0.75 V which is also present in solutions of $Mn_2^{III}OQ_4(HQ)_2$. The oxidation is irreversible in DMSO and pyridine but quasi-reversible in acetonitrile. Again, this appears to be a one-electron per manganese oxidation based on peak currents. Electrode filming by the oxidation product precluded confirmation by controlled potential coulometry.

When one equivalent of protons is added to a solution of $Mn^{II}Q_2$, both anodic peak currents are reduced by one half and a new cathodic peak appears at -1.85 V which corresponds to the reduction of the hydroxyl protons of free HQ (7). When another equivalent of protons is added, the anodic peaks disappear entirely and the only peak which remains is the cathodic peak at -1.85 V. Addition of OH$^-$ ions to solutions of $Mn^{II}Q_2$ also decreases the original anodic peak currents and a new anodic peak appears at +0.08 V which corresponds to the oxidation of free Q$^-$ ions to dimeric Q_2.

Addition of one equivalent of protons per mole of manganese atoms to solutions of $Mn_2^{III}OQ_4(HQ)_2$ slightly decreases the peak current for the reduction at -0.31 V and reveals a new small cathodic peak at +0.11 V. (The peak at +0.11 V is also observed as part of the reversible $Mn^{II}Q_2$ re-reduction following the oxidation at +0.16 V.) Addition of one equivalent of OH$^-$ ions per mole of manganese to solutions of $Mn_2^{III}OQ_4(HQ)_2$ results in a spontaneous chemical reduction, a decrease in the original cathodic peak at -0.31 V, and the formation of a new anodic peak at -0.16 V.

Reactions of Mn(II)- and Mn(III)-8-Quinolinol Complexes with O_2^-, O_2, and H_2O_2.

Figure 2 illustrates cyclic voltammograms in DMSO solution at a platinum electrode for a) 1.3 m$\underline{M}$ O_2^-, b) 1 m$\underline{M}$ $Mn^{II}Q_2 \cdot 2H_2O$, and c) a mixture of 1.3 m$\underline{M}$ O_2^- and 1 m$\underline{M}$ $Mn^{II}Q_2 \cdot 2H_2O$ ten seconds after mixing. Clearly all of the O_2^- is decomposed within the time it takes to record the first cyclic voltammogram, and the complex remains in its initial oxidation state and is not appreciably decomposed. The solution does not change color during the reaction. In addition there are cathodic peaks at -0.83 V and -1.30 V, which correspond to the reduction of O_2 and H_2O_2, respectively, and an enhancement of the anodic peak at -0.16 V. The relative yields of O_2 and H_2O_2 vary but are between 50 and 75% of the theoretical values. After standing for 30 minutes the cyclic voltammograms look similar to those obtained for solutions of pure $Mn_2^{III}OQ_4(HQ)_2$. At higher O_2^- to $Mn^{II}Q_2 \cdot 2H_2O$ concentration ratios there is 50 to 80% immediate decomposition, but then the rate of decomposition decreases. A large anodic peak appears at $+0.08$ V which corresponds to Q^- oxidation.

$Mn^{II}Q_3^-$ (prepared $\underline{in}$ $\underline{situ}$ by electrolysis of $Mn_2^{III}OQ_4(HQ)_2$ solutions at -0.50 V) and $Mn_2^{III}OQ_4(HQ)_2$ also react with O_2^- to form O_2 and H_2O_2 but at a slower rate than $Mn^{II}Q_2 \cdot 2H_2O$. In addition, $Mn^{II}Q_2 \cdot 2H_2O$ reacts slowly with both O_2 and H_2O_2 to yield solutions which have cyclic voltammograms similar to those of $Mn_2^{III}OQ_4(HQ)_2$. It is noteworthy that $Mn_2^{III}OQ_4(HQ)_2$ does not react with either O_2 or H_2O_2 in DMSO solution.

When 0.39 m$\underline{M}$ O_2^- and 1.25 m$\underline{M}$ $Mn^{II}Q_2 \cdot 2H_2O$ are mixed in pyridine the results are qualitatively similar to those obtained in DMSO solution. Hydrogen peroxide formation, however, is more clearly visible and the yields of O_2 and H_2O_2 are somewhat higher. At higher O_2^--to-$Mn^{II}Q_2 \cdot 2H_2O$ concentration ratios, although the decomposition of O_2^- remains rapid, the complex is destroyed as indicated by the absence of all peaks assignable to manganese species. Cyclic voltammograms recorded after the reaction show only O_2 and H_2O_2.

$Mn^{II}Q_2 \cdot 2H_2O$ also reacts slowly with H_2O_2 in pyridine to give solutions which after degassing with argon give cyclic voltammograms similar to $Mn^{III}OQ_4$-$(HQ)_2$. However, $Mn^{II}Q_2 \cdot 2H_2O$ reacts very rapidly with

O_2 in pyridine to give a solution with cyclic voltammograms devoid of Mn(II) or Mn(III) redox activity.
This indicates the formation of an insoluble or electrochemically inactive species.

The cyclic voltammetric experiments detailed in
Figure 2 were also carried out using $Mn^{II}Q_2$ and $Zn^{II}Q_2$
in place of $Mn^{II}Q_2 \cdot 2H_2O$. In both cases the rate of
decomposition for O_2^- is as slow as it is for DMSO
solutions without added metal complex (about 5-15% per
hour).

Discussion and Conclusions

Structure of the Complexes in Solution. The
conductance and molecular weight data indicate that
$Mn^{II}Q_2 \cdot H_2O$ in DMSO solution is involved in two
different equilibria; a dissociation reaction which
yields about 80% $Mn^{II}Q_2$ and H_2O, and a ligand transfer or hydrolysis which yields about 10% of some
charged species. The electrochemical data to be discussed later support the formulation of the charged
species as $Mn^{II}Q^+$ and $Mn^{II}Q_3$.

The manganese(III)-8-quinolinol complex also
undergoes structural changes upon dissolution into
DMSO as illustrated by its decrease in magnetic
moment. Because all known Mn(III)-DMSO complexes are
high-spin (9), solvent effects are unlikely to cause
spin pairing. The decreased value of the magnetic
moment implies some antiferromagnetic coupling of the
type that would be expected for μ-oxo bridged species
or di-μ-hydroxo bridged species, but is weaker than
that observed for di-μ-oxo bridged species (10,11).
The decrease in magnetic moment in the Mn(III)-8-
quinolinol complex is comparable to that observed for
μ-oxo-bis(tetraphenylporphinato)dimanganese(III,III)
where the solid state magnetic moment at 295 $^\circ K$ is
reported as 4.12 B.M. (12).

In contrast to the magnetic moment in DMSO solution, the near theoretical value for the mangan-
ese(III)-8-quinolinol complex in acetonitrile and pyridine solutions suggests monomeric structure in acetonitrile and pyridine.

The di-μ-oxo bridged manganese(IV) 1,10-phenan-
throline complex, $Mn_2^{IV}O_2(1,10\text{-phen})_4(ClO_4)_4$, has a
solid-state magnetic moment of 1.86 B.M. per manganese
ion (10). This value corresponds to one
unpaired electron per manganese ion rather than the
expected three unpaired electrons. Unfortunately, the
compound is not soluble enough to permit measurement
of its solution magnetic moment. The one-electron

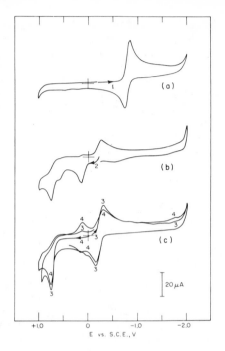

Figure 1. Cyclic voltammograms in 0.1M TEAP-DMSO at a Pt electrode of (a) 1.8mM O_2, (b) 1mM $Mn^{II}Q_2 \cdot 2H_2O$, (c) 1mM $Mn_2^{III}OQ_4(HQ)_2$. Scan rate, 0.1 V s⁻¹.

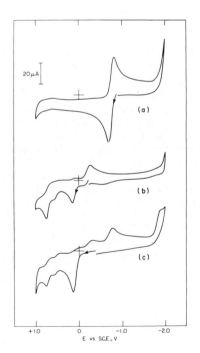

Figure 2. Cyclic voltammograms in 0.1M TEAP-DMSO at a Pt electrode of (a) 1.32mM O_2, (b) 100mM $Mn^{II}Q_2 \cdot 2H_2O$, and (c) a mixture of 1.32mM O_2^- and 1.00mM $Mn^{II}Q_2 \cdot 2H_2O$. Scan rate, 0.1 V s⁻¹.

reduction product of this dimer, $Mn_2^{III-IV}O_2(1,10-phen)_4^-$ $(ClO_4)_3$, is soluble, however, and has a magnetic moment of 1.56 B.M. per manganese ion in acetonitrile solution (_11_). This value corresponds to an average of less than one unpaired electron per manganese instead of the expected 3.5.
Similar results have been obtained for the di-μ-oxo manganese(III-IV) bipyridyl complex $Mn_2^{III-IV}O_2(bipy)_4-$ $(ClO_4)_3$. The magnetic moment in acetonitrile solution of 1.81 B.M. per manganese (_11_) is again close to the spin-only value of one unpaired electron per manganese. These results imply strong anti-ferromagnetic coupling across the dioxo bridge and confirm the stability of oxo-bridged species in solution.

The elemental analysis and solid-state magnetic moment of the manganese(III)-8-quinolinol complex imply that the complex exists as the monomeric tris chelate, $Mn^{III}Q_3 \cdot \frac{1}{2}H_2O \cdot CH_3OH$, in the solid state and becomes a bridged dimer in DMSO solution. There are several possible dimeric structures: a di-μ-oxo bridge, a μ-oxo bridge, or a di-μ-hydroxo bridge. The dioxo bridged structure can be ruled out because of the lack of strong antiferromagnetic coupling. The di-μ-hydroxy species, $Mn_2^{III}(OH)_2Q_4(HQ)_2$, and the μ-oxo species, $Mn_2^{III}OQ_4(HQ)_2$, are alternatives. Magnetic behavior for these species cannot be predicted because well characterized manganese(III) dimers with these types of bridging are unknown. The mono-oxo-bridged formulation is attractive, however, because the related iron(III)-8-quinolinol complex, $Fe^{III}Q_3 \cdot \frac{1}{2}H_2O$, becomes mono-oxo bridged in DMSO solution (_13_). Although the structure of the complex is still in doubt, the mono-oxo bridged formulation is reasonable and will be used throughout the remaining discussion.

Redox Properties of Mn(II)- and Mn(III)-8-Quinolinol Complexes.

A self-consistent redox mechanism for these compounds appears in Table I. The $Mn^{II}Q_2 \cdot 2H_2O$ complex is not reducible but is reversibly oxidized to $Mn^{III}Q_2^+$ at +0.16 V. Most of the oxidized species ends up as $Mn_2^{III}OQ_4(HQ)_2$. $Mn^{III}Q_2^+$ therefore must undergo a hydrolytic reaction with residual water to form $Mn^{III}Q_2(OH)$ and H^+. The H^+ ions that are formed attack unoxidized $Mn^{II}Q_2$ to form $Mn^{II}Q^+$ and HQ. This reaction has been confirmed by the experiments in which H^+ ion is added to solutions of $Mn^{II}Q_2$. The liberated HQ reacts with $Mn^{III}Q_2(OH)$ to form $Mn_2^{III}OQ_4(HQ)_2$ and H_2O. Furthermore, controlled-

Table I

Redox Reactions and Voltammetric Peak Potentials for Manganese(II) and –(III) 8-Quinolinol Complexes[a]

Reaction	E_{pc}	E_{pa}	e^-/Mn
	V vs. SCE		
$Mn^{II}Q_2 \cdot 2H_2O \longrightarrow Mn^{II}Q_2 + 2H_2O$ [b]			
$2Mn^{II}Q_2 \rightleftharpoons Mn^{II}Q^+ + Mn^{II}Q_3^-$ [c]			
$Mn^{II}Q_2 + Q^- \longrightarrow Mn^{II}Q_3^-$ [d]			
$Mn^{II}Q_3^- + H^+ \longrightarrow Mn^{II}Q_2 + HQ$			
$Mn^{II}Q_2 \rightleftharpoons Mn^{III}Q_2^+ + e^-$	+0.11	+0.16	1
$Mn^{III}Q_2^+ + H_2O \longrightarrow Mn^{III}Q_2(OH) + H^+$			
$Mn^{II}Q_2 + H^+ \rightleftharpoons Mn^{II}Q^+ + HQ$			
$2Mn^{III}Q_2(OH) + 2HQ \longrightarrow Mn_2^{III}OQ_4(HQ)_2 + H_2O$			
$Mn_2^{III}OQ_4(HQ)_2 + 2e^- \xrightarrow{DMSO} 2Mn^{II}Q_3^- + H_2O$	−0.31	−0.16	1
$2Mn^{III}Q_3(solid) + H_2O \rightleftharpoons Mn_2^{III}OQ_4(HQ)_2$			
$Mn_2^{III}OQ_4(HQ)_2 \longrightarrow Mn_2^{IV}OQ_4(HQ)_2^{2+} + 2e^-$	+0.70[e]	+0.75	
$Mn_2^{III}OQ_4(HQ)_2 + 2OH^- \longrightarrow Mn^{II}Q_3^- + Mn^{II}Q^+ + Q_2 + H_2O$			

[a] At a platinum electrode (scan rate, 0.1 V s^{-1}) in 0.1 $\underline{M}$ TEAP/DMSO.

[b] Cryoscopic molecular weight determination in DMSO indicates compound is 80% dissociated.

[c] Cyclic voltammetry indicates that 1 mM solutions of Mn$^{II}Q_2 \cdot 2H_2O$ in DMSO contain about 10% Mn$^{II}Q_3^-$, and conductance measurements indicate that such solutions dissociate about 10% into a 1-to-1 electrolyte.

[d] Q$^-$ was prepared electrochemically *in situ* by the reduction of HQ.

[e] Acetonitrile solution.

potential coulometry of $Mn^{II}Q_2$ at +0.25 V indicates that 0.5 equivalent of electrons is transferred per mole of complex instead of 1. This is consistent with the overall reaction

$$2 Mn^{II}Q_2 + \tfrac{1}{2}H_2O \longrightarrow \tfrac{1}{2}Mn_2^{III}OQ_4(HQ)_2 + Mn^{II}Q^+ + 1e^- \quad (5)$$

$Mn_2^{III}OQ_4(HQ)_2$ is reduced, in the absence of acid, by a one electron-per-manganese ion process at -0.31 V to form $Mn^{II}Q_3^-$ and H_2O. The electron stoichiometry has been confirmed by controlled potential coulometry at -0.50 V. The formulation of the reduced species as $Mn^{II}Q_3^-$ is supported by experiments in which Q^- ion is electrochemically generated in situ in the presence of $Mn^{II}Q_2$. The first anodic peak shifts from +0.16 to -0.16 V, the potential at which the species formed after reduction of $Mn_2^{III}OQ_4(HQ)_2$ is reoxidized. When H^+ ions are present, $Mn^{II}Q_3^-$ is converted to $Mn^{II}Q_2$ and HQ. The anodic peak corresponding to the reoxidation of the product species shifts back to +0.16 V and a cathodic peak appears at -1.85 V due to HQ reduction.

The $Mn_2^{III}OQ_4(HQ)_2$ complex is irreversibly oxidized at +0.75 V to form a manganese(IV) species which immediately oxidizes the solvent, the residual water present in the solvent, or its own ligands, and forms a mixture of the two manganese(III) complexes again. Finally, $Mn_2^{III}OQ_4(HQ)_2$ is chemically reduced in the presence of OH^- ions to give a species which is oxidized at -0.16 V. This potential corresponds to the oxidation of $Mn^{II}Q_3^-$. The reducing agent is probably Q^- which is oxidized to Q_2. The $Mn^{II}Q_2$ complex also is highly susceptible to nucleophilic displacement of the Q^- ligands by OH^- ions. In addition, $Mn^{II}Q_2$ reacts slowly with both O_2 and H_2O_2 and, based on the peak potentials, the product species of the reaction is $Mn_2^{III}OQ_4(HQ)_2$. The extra HQ ligands must of necessity come from $Mn^{II}Q_2$ so the other product of the reaction is $Mn^{II}Q^+$.

Reaction of the Complexes with O_2^-. The manganese(II)-8-quinolinol complexes, $Mn^{II}Q_2 \cdot 2H_2O$ and $Mn^{II}Q_3$, and the manganese(III)-8-quinolinol complex, $Mn_2^{III}OQ_4(HQ)_2 \cdot 2CH_3OH$, in DMSO solution represent a system that undergoes oxidation-reduction chemistry which parallels much of that observed for mitochondrial superoxide dismutase. $Mn^{II}Q_2$ accelerates the decomposition of O_2^- to form nearly stoichiometric amounts of the correct products, O_2 and a mixture of

H_2O_2 and HO_2^-. The increased peak current that appears to be associated with the oxidation of $Mn^{II}Q_2$ at +0.16 V probably is due to the coincident oxidation of the decomposition product, HO_2^- to O_2. When the scan is reversed after the anodic peak at +0.16 V the reduction peak at -0.83 V is enhanced. The cyclic voltammograms and the solution color indicate that the complex itself is in its initial oxidation state and is not appreciably decomposed.

The catalytic properties of the system are complicated by the fact that the only source of protons for HO_2^- and H_2O_2 formation is H_2O. The OH^- ions thus generated attack the catalyst to form what appears to be an inactive species. When the catalyst is used at lower concentrations (0.5 mM or less), the decomposition of O_2^- is rapid at first but slows down quickly, presumably because of base-induced decomposition of the catalyst, and also because the increasing basicity of the medium decreases the proton activity, which in turn inhibits peroxide formation. In addition, the O_2 and H_2O_2 which are formed react with $Mn^{II}Q_2$ to form $Mn_2^{III}OQ_4(HQ)_2$.

$Mn_2^{III}OQ_4(HQ)_2$ and $Mn^{II}Q_3^-$ also react with O_2^- to give O_2 and H_2O_2, although more slowly. Based on our observations, a reasonable mechanism for the $Mn^{II}Q_2$ catalyzed disproportionation of O_2^- appears to be

$$Mn^{II}Q_2(H_2O)_2 + O_2^- \longrightarrow Mn^{III}Q_2(O_2H)(H_2O) + OH^- \quad (6)$$

$$Mn^{III}Q_2(O_2H)(H_2O) + O_2^- \xrightarrow{H_2O} Mn^{II}Q_2(H_2O)_2 + O_2 + HO_2^- \quad (7)$$

with secondary reactions

$$Mn^{II}Q_3^- + O_2^- \longrightarrow Mn^{III}Q_3(O_2H)^- + OH^- \quad (8)$$

$$Mn^{III}Q_3(O_2H)^- + O_2^- \longrightarrow Mn^{II}Q_3^- + O_2 + HO_2^- \quad (9)$$

The most remarkable feature of this model is the apparent ability of a manganese(II) complex to reduce O_2^- to HO_2^-. Such a process, on the basis of the electrochemical peak potentials for the individual components, Figure 1, appears to be thermodynamically impossible. Apparently Equation 5 is favored as a result of the strong interaction of the product species, Mn^{III} and HO_2^-.

Several unsuccessful attempts have been made to isolate a manganese(III) 8-quinolinol-peroxide complex. The following two reactions in DMSO with the indicated reaction stoichiometries

$$6Mn^{II}Q_2 + 2O_2 + 3H_2O \longrightarrow Mn_2^{III}OQ_4(HQ)_2 +$$
$$2Mn^{III}Q_2(HO_2) + 2Mn^{II}Q(OH) \quad (10)$$

$$Mn^{II}Q_2 + H_2O_2 \xrightarrow{+0.25V} Mn^{III}Q_2(HO_2) + H^+ + e^- \quad (11)$$

were studied by cyclic voltammetry. Both solutions exhibit an additional cathodic peak at -0.75 V (besides the peak at -0.31 V for $Mn_2^{III}OQ_4(HQ_2)$ which may result from the reduction of $Mn^{III}Q_2(HO_2)$ or of bound peroxide ions.

The apparent oxidation by O_2^- ion (generated from oxygen plus xanthine-xanthine oxidase or from illumination of spinach chloroplasts) of Mn(II) to Mn(III) in the presence of a pyrophosphate buffer system was reported recently (14). However, the authors conclude that the manganese(II)-pyrophosphate complex does not act as a disproportionation catalyst for O_2^- ions, and imply that the manganese(III)-pyrophosphate does not oxidize O_2^- ion but does oxidize H_2O_2. This is in sharp contrast to the results of the present study (Equations 6-9).

Although the Fridovich mechanism (Equations 1-4) invokes a third oxidation state for the enzyme to rationalize the kinetic results, there is another possible explanation. The less catalytic state of the enzyme may simply be a different hydrolytic species.

Because protons also catalyze the disproportionation of O_2^- ions and coordinated DMSO might serve as a proton source, the cyclic voltammetric experiments detailed in Figure 2 have been made in pyridine. The results are qualitatively the same. Combination of 1.25 m$\underline{M}$ $Mn^{II}Q_2$ with 0.39 m$\underline{M}$ O_2^- immediately destroys all of the O_2^- and yields a solution that contains $Mn^{II}Q_2$, O_2, and H_2O_2 in amounts equivalent to 70-90% efficiency for Equations 6-9. At higher O_2^--to-$Mn^{II}Q_2 \cdot 2H_2O$ ratios the complex is destroyed. The fact that the complex is more easily destroyed in pyridine solution during the catalytic reaction can be explained on the basis of the greatly accelerated reaction of one of the products, O_2, with $Mn^{II}Q_2$. Since separate experiments have shown that O_2 reacts rapidly with $Mn^{II}Q_2$ in pyridine to form an electrochemically inactive species, it is expected that the catalyst will disappear when the O_2^- to $Mn^{II}Q_2$ concentration ratio becomes higher, and thus the catalytically generated oxygen concentration increases.

Another possible interpretation of the results is that the $Mn^{II}Q_2$ complex acts as a Lewis acid to catalyze the disproportionation of O_2^- ions. Charge neutralization of one O_2^- ion by the Lewis acid would make it electrostatically easier for O_2^- collisions and electron transfer to occur. However, when the experiment summarized in Figure 2 is repeated with $Mg^{II}Q_2$ or $Zn^{II}Q_2$ substituted for $Mn^{II}Q_2$, the rate of decomposition for O_2^- is not any faster than the normal decomposition rate in DMSO. This supports the conclusion that $Mn^{II}Q_2$ acts as a redox catalyst.

The present study is being extended to determine what effects the ligands have on both the redox chemistry of the complexes and their ability to catalyze the decomposition of superoxide ion. In particular, ligands are being sought which will stabilize manganese(II) complexes toward nucleophilic displacement by OH^- ions and which will accelerate the reaction in Equation 6. We also are trying to find a suitable buffer system to facilitate the formation of hydrogen peroxide and prevent the ultimate destruction of the catalyst. Such a system will make it possible to evaluate the kinetic parameters for the various reactions.

Acknowledgment

This work was supported by the U. S. Public Health Service-NIH under Grant No. GM 22761.

Literature Cited

1. J. M. McCord and I. Fridovich, J. Biol. Chem. (1969), 244, 6049.
2. B. B. Keele, Jr., J. M. McCord, and I. Fridovich, J. Biol. Chem. (1970), 245, 6176.
3. P. G. Vance, B. B. Keele, Jr., and K. V. Rajagopalan, J. Biol. Chem. (1972), 247, 4782.
4. R. A. Weisiger and I. Fridovich, J. Biol. Chem. (1973), 248, 3582.
5. M. Pick, J. Rabani, F. Yost, and I. Fridovich, J. Am. Chem. Soc. (1974), 96, 7329.
6. A. D. Goolsby and D. T. Sawyer, Anal. Chem. (1967), 39, 411.
7. A. F. Isbell, Jr., and D. T. Sawyer, Inorg. Chem. (1971), 10, 2449.
8. D. F. Evans, J. Chem. Soc. (1959), 2003.
9. C. P. Prabhakaran and C. C. Pratel, J. Inorg. Nucl. Chem. (1968), 30, 867.

10. H. A. Goodwin and R. N. Sylva, Aust. J. Chem. (1967), 20, 629.

11. M. M. Morrison and D. T. Sawyer, submitted to J. Am. Chem. Soc. (1976).

12. E. B. Fleischer, J. M. Palmer, T. S. Srivastava, and A. Chatterjee, J. Am. Chem. Soc. (1971), 93, 3162.

13. T. L. Riechel and D. T. Sawyer, submitted to Inorganic Chemistry (1976).

14. T. Kono, M. Takahashi, and K. Asada, Arch. Biochem. Biophys. (1976), 174, 454.

Interfacial Behavior of Biologically Important Purines at the Mercury Solution Interface

H. KINOSHITA, S. D. CHRISTIAN, M. H. KIM,
J. G. BAKER, and GLENN DRYHURST

Department of Chemistry, University of Oklahoma, Norman, Okla. 73019

Over the past several years an increasing body of evidence has suggested that interactions of nucleic acids with electrically charged membranes play an important role in many basic biological processes. However, the physics and chemistry of biodynamic molecules adsorbed at membrane-fluid interfaces in living organisms are not at all well understood. Basic studies of interfacial phenomena in vivo are rendered difficult by the multitude of surface active compounds which exist in biological fluids and the virtual impossibility of identifying specific interactions between a given adsorbate molecule and active sites on a biological interface.

Electric fields have been known for some time to influence the conformation of various natural and biosynthetic polynucleotides in solution. For example, Hill (1) has calculated that high electric fields could bring about separation of the two molecular chains of nucleotides in DNA. Based on birefringence measurements, it has been demonstrated (2) that in high intensity electrical fields ($>10^4$ Vcm^{-1}) DNA first aggregates and then undergoes a structural transition in which the angles of the purine and pyrimidine bases with respect to the helix axis are altered. In electric fields of about 2×10^4 Vcm^{-1} ribosomal RNA and polynucleotides such as poly(A).2 poly(U) appear to undergo a transient opening of base pairs followed by only partial reassociation of the unfolded regions (3). Such structural effects of electrical fields on natural polynucleotides have been implicated in the mechanisms of nerve impulse transmission and information storage in the central nervous system (2), perhaps as an initial step in the recording of biological memory (3-6).

The potentials that exist at certain biological membranes such as a cell membrane are thought to be of the order of 0.1V. In a biological fluid having an ionic strength of 0.1 to 0.2 this potential would extend over distances of $10-100\mathring{A}$ (7). This corresponds to electric fields of $10^5-10^6 Vcm^{-1}$. Clearly, if a biopolymer such as DNA or RNA is present in close proximity to such a biosurface then it seems quite reasonable to suggest that macromolecular structural transitions might occur. In fact, in living organisms DNA, for example, is partially associated with the nuclear or cytoplasmic membrane, or with the interface of the nucleolus (8-11). A theory has been advanced that replication could begin at the level of the nuclear or cellular wall (10-12). Indeed, Hill (1) has suggested that the electric fields and their variations at biological interfaces might act as the trigger for division of genetic material in the cell prior to self-duplication.

A very interesting aspect of electrical activity associated with biological processes is the existence of the potential of injury at a trauma site. A significant observation is that the injury potential follows a different time course in the healing of, for example, a limb amputation in the case of a species which can regenerate the limb as opposed to one which exhibits only scar formation (13, 14). It has been found that implantation of small electrodes at the injury site in a nonregenerating species (forcing the injury potential to approximate that of a regenerating species) causes at least partial limb regeneration even in a complex species such as the rat (15,16). This implies that fundamental biological processes (ultimately at the genetic level) may be controlled by the natural or artificially applied electrical environment at a tissue repair site. The application of small, locally applied electric fields has been used for the stimulation of bone healing as a very practical outcome of such studies.

Cope (17-24) has presented some convincing arguments that a cell surface-biological fluid interface may be regarded as being very similar to a liquid-solid interface which exhibits electrical behavior analogous to that occurring at an electrode-solution interface.

An electrode-solution interface is generally characterized by a well-defined electrical double layer bounded on one side by the electrode surface and on the other side by an ionic layer across which a relatively high electrical field develops (up to

ca. 10^6V cm^{-1}). This constitutes the so-called
inner or compact double layer, the width of which is
only a few atomic diameters. On the boundary between
the inner and the diffuse double layers the field
strength has only about 1/10 of its original value,
and it then decreases in the diffuse double layer to
virtually zero. In a medium of ionic strength ca.
0.1-0.2 (typical of biological fluids) the depth of
the diffuse double layer is about 100Å (25). An exactly similarly structured region exists at a cell
membrane-biological fluid interface (26).

At physiological bulk-phase pH value of 7.2 all
mammalian cells so far examined carry a net negative
charge at their surfaces. However, the surface potential of a cell is not constant but can undergo
some rather dramatic changes. For example, cells
isolated from the regenerating livers of rats some
days after partial hepatectomy and cells from neo-
nates have significantly higher electrophoretic mobilities than liver cells from normal adults (27),
i.e., cell proliferation is associated with increased
net surface negativity. Similarly, the electrophoretic mobilities of certain tumor cells increase with
growth rate (28). At the time of mitosis a very significant increase in net surface negativity has been
observed in various types of cultured tumor cells
(29, 30). Indeed, Ambrose et al. (31, 32), have noted
a correlation between malignancy and increased cell
surface negativity, although this is certainly not
thought to be a universal correlation (33).

A substantial amount of evidence is being developed which indicates that interactions with biological interfaces is a prerequisite for the manifestation of the biological effects of polynucleotides in
mammalian cell systems *in vivo* and *in vitro*. Thus,
Field et al. (34) have reported that RNA double
strands, and particularly poly(I).poly(C) induce interferon formation in mammalian cells. Subsequently,
Schell (35) has shown that poly(I).poly(C) is adsorbed to the outside of the cell followed by strand
separation and ultimately by interferon formation.
It has further been suggested (36) that other biological effects of polynucleotides, such as adjuvant
effects and enzyme inhibition/activation requires interaction of the polynucleotide with the charged cell
surface.

Thus, in summary, the surfaces of mammalian
cells and other biological membranes carry an appreciable electrical potential. The electrical
double-layer formed in the immediate vicinity of a

charged membrane-biological fluid interface is essentially identical to that formed at an electrode surface. The potential of a cell membrane surface is known to alter when processes such as cell regeneration occur or, often, when a cell becomes malignant and particularly at the time of mitosis. The potential of injury at a site of trauma is clearly implicated in control of the tissue regeneration process.

The electrical fields in the immediate vicinity of a cell surface are probably very large (10^4-10^6 Vcm^{-1}) although they extend over only very small distances ($\underline{ca}$. 10-100 Å). Such intense electric fields have been shown to cause structural transitions in certain natural and biosynthetic polynucleotides. In addition, the interaction (adsorption) of polynucleotides at the charged cell surface appears to be a prerequisite for manifestation of the biological effects of the polynucleotides.

Because a charged cell surface-biological fluid interface is similar to a charged electrode-electrolyte solution interface, it seems reasonable that an understanding of the interfacial behavior of biomolecules at the latter interface might reveal significant information regarding the interactions of these molecules at biological interfaces.

It would seem to be self-evident, however, that interfacial studies of nucleic acids and other polynucleotides, and interpretation of the data so collected, must rely on a fundamental knowledge of the interfacial behavior of the monomeric units, i.e., bases, nucleosides and nucleotides. A number of investigators have reported that various monomeric purine and pyrimidine derivatives are adsorbed at mercury electrodes (37-45). Such studies, however, have generally been very qualitative; they have revealed virtually nothing about the surface areas occupied by the adsorbed molecules and hence their probable surface orientations, the nature of the adsorption isotherms and the effects of potential on the adsorption processes, the intermolecular interactions between adsorbed molecules and interactions between the adsorbed molecules and the electrode surface.

Recently, Nürnberg et al. (46) have reported on the adsorption of adenosine and adenosine mononucleotides at a mercury electrode. Measurements of the amount of the adenine species adsorbed were accomplished by use of the time integral of the faradaic reduction peak of these molecules at a stationary mercury drop electrode at 5°C and at pH 3.4. This

voltammetric technique, however, is subject to many
experimental limitations and problems (47).
We have begun a systematic investigation of the
interfacial behavior of the purine and pyrimidine
bases, deoxynucleosides and deoxynucleotides found in
nucleic acids. In this report the interfacial behavior of various adenine derivatives at pH 9 at mercury
electrodes will be described. Two major surface electrochemical techniques were employed in these studies
-- the capillary electrometer and differential capacitance measurements. Some preliminary results using
electroellipsometry will also be presented.

Experimental

        Differential capacitance measurements were
obtained by a phase-selective a.c. polarographic
method. A Princeton Applied Research Corporation
(PARC) Model 174 Polarographic Analyzer coupled with
a PARC Model 174/50 AC Polarographic Analyzer Interface and a PARC Model 121 Lock-in Amplifier/Phase
Detector were employed for differential capacitance
measurements. A phase angle of 90° with respect to
the applied alternating voltage was employed. At the
pH values employed in this study, adenine and its derivatives are not electrochemically reducible. The
dropping mercury electrode (DME) was siliconized (39)
and was equipped with a mechanical drop dislodger.
A pool of mercury inserted at the bottom of a thermo-
statted 5ml capacity cell served as the counter electrode. A saturated calomel reference electrode (SCE)
was employed using a fine Luggin capillary positioned
close to the tip of the DME. A.c. polarograms were
usually obtained at a frequency of 100Hz and with
a modulating amplitude of 10mV peak-to-peak. Thus,
the capacity results reported here were all measured
at a frequency of 100Hz and were not extrapolated to
zero frequency. However, the capacity values were
virtually independent of frequency between about 50
and 600 Hz. All measurements were made without damp-
ing on the Model 174 and using a controlled droptime
of 2.00s. When the a.c. polarogram was recorded on
an X-Y recorder (Hewlett-Packard Model 7001A) the
d.c. potential was scanned at a sweep rate of
$0.005Vs^{-1}$. However, in some of our later studies
the alternating current was measured by use of a
Keithley Model 168 Autoranging Digital Multimeter
connected to the Y axis (current) output of the PARC
Model 174.
        The capillary electrometer and its associated

pressure system is shown schematically in Figure 1.
The Pyrex capillaries had a diameter, 1mm above the
tip, of 0.002-0.005mm. The capillaries were normally
aged for several days by soaking in deionized water
and subsequently stored with the tip immersed in water.
A Brinkmann/Wenking Model LT73 potentiostat was used.
The location of the mercury column in the capillary
was observed with a Gaertner 2206-A Cathetometer
through an optically flat Pyrex window sealed on one
side of the cell (Figure 1). The cell was water
jacketed and maintained at a temperature of 25±0.1°C.
The pressure at the mercury-test solution interface
was varied by applying pressure to the gas above the
mercury by means of a syringe and two micrometer
burets. The coarse adjust utilized a 20ml plastic
syringe while fine pressure adjustment was accomplished with two Gilmont 2.0ml Micrometer Burets. A
Mensor Corporation Quartz Manometer pressure gauge was
used to measure the gas pressure.

Test solutions were deaerated with nitrogen for
at least 15 minutes before measurements were carried
out. All potential measurements utilized a saturated
calomel reference electrode at 25°C.

Data points were taken at 50mV intervals from
-0.2V to -1.8V. A drop of mercury was expelled from
the capillary before the electrocapillary curve at
each concentration was measured. The pressure at
each applied potential was then adjusted to bring
the mercury to the reference point in the capillary
(1mm from the tip). The height of the mercury column was measured with the Gaertner cathetometer to a
precision of 0.02mm before and after measurement of
each electrocapillary curve. A correction was applied to the measured pressure for this height and
for the small back pressure of the solution.

The borate buffer pH 9 utilized was constituted
as follows: 17.5g $Na_2B_4O_7.10H_2O$, 67.7g KCl and 16.85
ml 1M HCl diluted to 1l with deionized water.
McIlvaine buffer pH 7 was constituted as follows:
58.9g $Na_2HPO_4.12H_2O$, 3.7g citric acid.H_2O and 5.4g
KCl diluted to 1l with water. Both of these buffer
solutions have an ionic strength of 0.5. Sample
solutions were prepared by dissolving the adenine
species in the appropriate volume of buffer.

A Gaertner Model L119 Ellipsometer was employed
for ellipsometric studies. A schematic diagram of
the apparatus utilized is shown in Figure 2. The
angle of incidence of the light beam was 70°. Test
solutions were deaerated for about 30 minutes and a
nitrogen atmosphere was maintained over the solution

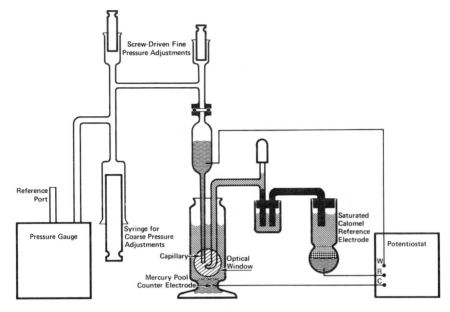

Figure 1. System for electrocapillary measurements

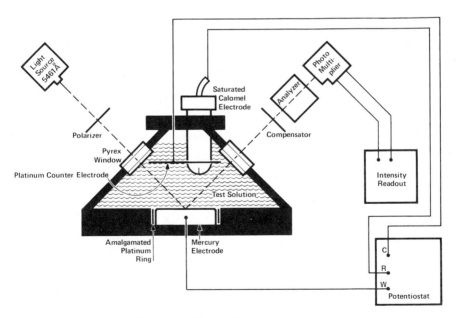

Figure 2. Electroellipsometry system

when measurements were taken. The circular mercury
pool working electrode had an area of 5.07cm^2. The
curvature of the mercury surface was reduced by
placing an amalgamated platinum insert around the
periphery of the pool. A small cross-sectional area
light beam, 0.8mm^2, was used to minimize the effect of
curvature of the mercury electrode surface. The procedure used in measuring thicknesses and changes in
thickness of films with the ellipsometer was the following. First, with only the background electrolyte
(buffer) solution in the cell, values of the polarizer
angle (P) producing a minimum in intensity of the reflected beam were determined as a function of potential in the range -0.2 to -1.6 V vs. SCE. (The analyzer angle, A, was not varied during these experiments
since the optimum value of A is only slightly affected
by the presence of films less than 10 or 20 Angstroms
thick and since the choice of analyzer angle in the
minimum region does not influence the value of P lead-
ing to minimum intensity.) Next, with a solution of
adsorbate in the cell, new values of P producing minimum intensity were determined. Values of film thickness were calculated from changes in P (at fixed A and
fixed potential) by using a computer program developed
by McCrackin (63).

In practice, it is sometimes difficult to repro-
duce exactly the values of P which produce minimum
intensity for a given adsorbate or background solution. However, the shapes of the various P vs. potential curves are ordinarily quite reproducible, and
it is possible to use the extensive adsorption results available from the capacitance and electrocapillary experiments to "normalize" the ellipsometric data. Thus, simple vertical displacement of the
observed P vs. potential curves usually brings them
into coincidence with the background curve in regions
of potential where little or no adsorption occurs.
Differences between the displaced curves for the adsorbate solutions and the background curve can then
be attributed directly to the optical effect of the
film and interpreted to yield film thicknesses as a
function of adsorbate concentration and potential.

Analysis of Capacitance and Electrocapillary Data

Using the a.c. polarographic method the capacitance, μFcm^{-2}, is readily determined from the observed
alternating current, μA, from the equation:

$$C = \frac{I}{\Delta E A 2 \pi f}$$

(1)

where ΔE is the amplitude of the applied alternating voltage (V), A is the surface area of the DME (cm^2) at the time the current is sampled, and f is the frequency (Hz) of applied alternating voltage. This equation is valid when the resistance of the test solution is small and the frequency of the applied alternating voltage is low (48).

The general procedure involved in processing capacitance data to obtain adsorption isotherms involved first, calculation of surface spreading pressure values, π, for the organic compound at the mercury-solution interface by back integration of capacitance data. Then, from values of π at known potential, E, and concentrations or activities, the parameters in the Frumkin adsorption equation were calculated by a nonlinear least squares analysis.

The back integration method (49) relies on the assumption that at sufficiently negative potentials, capacitance versus potential curves (C vs. E) for aqueous solutions of an organic adsorbate coincide with the background C vs. E curve for the electrolyte alone. This condition was met for all the adenine systems described here, i.e., at potentials between -1.6V and -1.8V the C vs. E curves become coincident and remain so at even more negative potentials. Integration of C vs. E curves for the background electrolyte solution (i.e., C_w vs. E) gives values of charge q_w, relative q*, the charge at the mercury-electrolyte solution interface at the starting potential E* (this was typically -1.8V in these studies). Thus,

$$q_w - q^* = \int_{E^*}^{E} CdE \qquad (\underline{2}).$$

By measurement of an electrocapillary curve of interfacial tension versus potential (γ_w vs. E) on the background electrolyte solution the value of the electrocapillary maximum potential (ECM) may be obtained (i.e., E at γ max.). At the ECM $q_w=0$, hence the absolute value of q_w at any potential E ($q_w(E)$) is

$$q_w(E) = [q_w(E) - q^*] - [q_w(ECM) - q^*] (\underline{3}).$$

From the measured values of $\gamma_w(E)$ determined from electrocapillary curves of the background solution the value of $q_w(E)$ may be determined by differentiation of γ values, i.e.,

$$q_w(E) = \frac{d\gamma_w}{dE} \qquad (\underline{4}).$$

Charge ($q_w(E)$) values determined by electro-capillary and capacitance methods always agreed well.

Capacitance measurement on solutions of known activity of the organic compounds were back integrated from E* in exactly the same fashion. Assuming at E* that $q_w = q_{org}$, then

$$q_{org}(E) = [q_{org}(E) - q^*] + q_w(E^*) \qquad (\underline{5}).$$

Thus, absolute values of $q_{org}(E)$ may be obtained. These values again agreed well with $q_{org}(E)$ obtained by differentiation of electrocapillary data (γ_{org} vs. E) for solutions of the organic species.

To obtain values of the interfacial tension ($\gamma(E)$) it is necessary to perform a second integration. We can write

$$\gamma - \gamma^* = -\int_{E^*}^{E} q \, dE \qquad (\underline{6}).$$

Now, γ^* (the interfacial tension at the starting potential) is known from electrocapillary measurements on the background electrolyte solution; moreover, it does not change with addition of adsorbate to the solution. Therefore, eq (6) can be used to obtain the interfacial tension as a function of both E and a, the activity of the organic solute. The spreading pressure at any solute activity and potential is

$$\pi = \gamma_w(E) - \gamma(E) \qquad (\underline{7}),$$

where γ_w is the value of γ for the background electrolyte solution at a = 0. With the exception of the deoxyadenosine monophosphate system (vide infra), the assumption is made throughout that the solute activity can be replaced by its molar concentration in the dilute solutions employed in the study.

The π values may be measured directly from electrocapillary data using eq (7).

Using this approach, therefore, it is possible to compare charge, q, and surface spreading pressure values, π, at any potential and concentration derived from capacitance and electrocapillary data.

Within the accuracy of our measurements, plots of π vs. ℓn a for all compounds reported here at potentials more negative than ca. -0.5V were superimposable by simple horizontal translation. This implies that the isotherms are congruent with respect to potential and that the attraction coefficient, α, in the Frumkin adsorption equation (50) is independent of potential (vide infra).

The method used to fit data to the Frumkin iso-
therm equation

$$\frac{\Theta}{1-\Theta} = Ba\ e^{2\alpha\Theta} \qquad (8)$$

is similar to that employed by Hansen and coworkers
(51). In this fixed potential form of the Frumkin
equation, Θ is the fraction of the electrode surface
covered with monolayer adsorbate, B and α are constants depending on the interactions between the adsorbed molecules and the surface and on lateral
intermolecular interactions between the adsorbate
molecules, respectively. The constant B is dependent
on potential according to eq (9) (52).

$$B = B_o e^{(-\Phi/\Gamma_m RT)} \qquad (9)$$

where Γm is the limiting surface excess of the solute
at full monolayer coverage in moles per unit area.
The term B_o is the value of B at the electrocapillary
maximum potential for the electrolyte, i.e., at a=0.
Combining equations (8) and (9) an equation referred
to as the generalized Frumkin equation is obtained
(eq 10).

$$\frac{\Theta}{1-\Theta} = B_o ae^{(-\Phi/\Gamma_m RT)} \cdot e^{2\alpha\Theta} \qquad (10)$$

In this equation the function Φ is given by

$$\Phi = \int_o^E q_w dE + C'E\ (E_N - 1/2\ E) \qquad (11)$$

where q_w is the charge at a potential E for background
electrolyte solution, C' is the capacitance of the
electrode completely covered with adsorbate monolayer
(assumed constant) (53) and E_N is the electrocapillary
maximum potential for the mercury-solution interface
at $\Theta=1$ measured relative to the ECM for the electrolyte alone. The potential, E, in eq (11) is also
relative to the ECM for electrolyte solution. Integration of eq (11) gives

$$\Phi = G(E) + C'EE_N - \frac{C'E^2}{2} \qquad (12)$$

where $G(E) = \gamma_w\ (O) - \gamma_w\ (E) \qquad (13)$.

Hence G(E) may be determined from the difference between the interfacial tension at the ECM and the value at any potential E (vs. ECM at a=0). This was normally obtained directly from electrocapillary curves on the electrolyte solution.

The Gibbs adsorption equation

$$\Gamma = \frac{-1}{RT} \frac{d\pi}{d\ln a} \tag{14}$$

may be combined with the Frumkin equation (eq [8]) and integrated to give an expression that relates π to α, Θ and Γ_m (51).

$$\pi = \Gamma_m RT \; [-\ln(1-\Theta)-\alpha\Theta^2] \tag{15}.$$

We have developed a nonlinear least squares method to fit π values at various concentrations and potentials to give the best values of B_o, α, C', Γ_m and E_N (54). This is done by first taking trial values of the latter 5 parameters and calculating values of Θ from eq (10) for each pair of a_i and E_i values by an iterative numerical method. This set of calculated Θ_i values is then used to predict a set of π_i values using eq (15). For a given set of parameters, a value of the sum of squares of the residuals s = $\Sigma_i (\pi_i - \pi \text{ calculated})^2$ is then obtained. The value of s is minimized with respect to variation of all five parameters; standard errors in all the constants and the root mean square deviation in π are also calculated. For most of the systems, convergence was obtained within 5 to 20 iteration cycles provided reasonably good initial estimates of the parameter values were made. In those systems where electrocapillary and capacitance data were available, it was often useful to fix the E_N value, which is estimated readily from electrocapillary curves for large activities of the organic adsorbate (i.e., at $\Theta \rightarrow 1$). In addition, C' may generally be estimated by measuring the capacitance of a nearly saturated solution of the organic compound at a potential at, or very close to, the potential of maximum adsorption (this is easily recognized as the potential at which π reaches its maximum value in π vs. E plots particularly in solutions where $\Theta \rightarrow 1$). The latter approach has also been used by Hansen and co-workers (55).

Results and Discussion

In this study, the interfacial behavior of ade-

nine and its deoxynucleoside and monodeoxynucleotides
was studied at pH 9 in borate buffer by use of differential capacitance and electrocapillary studies.
In pH 9 borate buffer the C vs. E curves of background
electrolyte and all solutions containing organic adsorbate become coincident at negative potentials,
which is a necessary condition if the back-integration
of capacitance method of Grahame et al. (49) is used.

Capacitance and Electrocapillary Studies

A typical series of capacitance curves for adenine is shown in Figure 3 where it may be seen that
up to a concentration of ca. 2-3mM a systematic decrease of the capacitance occurs at around -0.3 to
-0.7V with a broad desorption peak centered around
-1.2V. At concentrations at or above 4mM a sudden
sharp decrease in capacitance is observed centered
at -0.6V giving rise to a very sharply defined pit or
well. A similar pit is observed in the case of deoxyadenosine (Figure 4) although for this compound the
pit is centered at -1.25V. The monodeoxynucleotides
do not exhibit a pit at pH 9. Other workers have
observed similar pits at lower pH for many of these
types of compounds (40-44). Interpretation of capacitance results by the methods used in this study was
not possible in the pit region. However, electrocapillary measurements could be interpreted within the
pit region.

The adsorption processes occurring at concentrations below those necessary for pit formation will be
referred to as formation of the first adsorption
layer.

Using the calculational approach outlined earlier
surface spreading pressure values, π, versus the logarithm of activity (concentration) of adenine and its
deoxynucleoside and three monodeoxynucleotides at
various electrode potentials are readily superimposable by abscissa translation (Figure 5). The precision of the data, expressed as root mean square (rms)
deviations in π in Figure 5 is somewhat poorer than
results for some simple aliphatic compounds reported
by Baikerikar and Hansen (56), who employed a similar
computational method. This is probably due to the
greater complexity of the molecules studied here and,
perhaps, the slightly lower precision of the a.c.
polarographic measurement of capacitance used here
compared to the more conventional bridge methods for
capacitance measurements. Nevertheless, the fact
that the π vs. ℓn a plots are superimposable implies

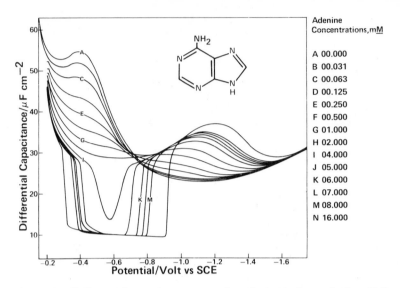

Figure 3.   Differential capacitance curves for adenine in borate buffer pH 9 at the DME. Curves obtained at 100Hz and 10mV, peak-to-peak.

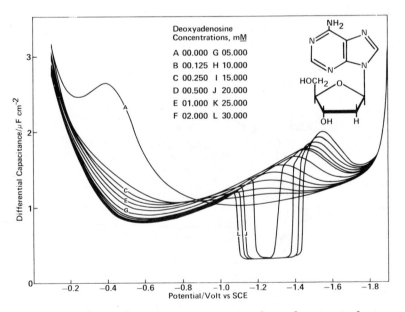

Figure 4.   Differential capacitance curves for deoxyadenosine in borate buffer pH 9 (ionic strength 0.5) at the DME. Curves obtained at 100Hz and 10mV peak-to-peak.

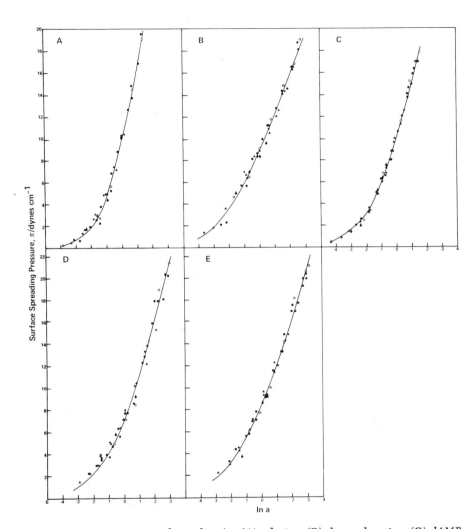

Figure 5. Composite π vs. ln a plots for (A) adenine, (B) deoxyadenosine, (C) dAMP, (D) dADP, and (E) dATP at pH 9 in borate buffer. Data points are for potentials from −1.0V to −0.4V. The rms deviation in π from the calculated curve for (A) is 0.54, (B) 1.27, (C) 0.31, (D) 0.79, and (E) 0.46 dyne cm⁻¹. The calculated curve for dAMP (C) takes into account the self-association of this compound using a sequential equilibrium model and a value of the equilibrium constant of 110 l/mol.

that for molecules in this adenine series the intermolecular interactions between adsorbate molecules on
the electrode surface are essentially independent of
the electrode potential (i.e., α in eq (8) is independent of potential).

The Frumkin adsorption model (with constant α)
employed in fitting integrated capacitance data requires that the linear relationship between charge,
q, and surface coverage (θ) be valid at any fixed potential according (57) to eq (16) where q' is the
charge at a

$$q = q_w(1-\theta) + q'(\theta) \qquad (16)$$

potential, E, corresponding to a completely covered
monolayer of organic adsorbate (i.e., the film at $\theta=1$).
This equation was tested by plotting q vs. values of
θ determined from the nonlinear least squares analysis
of π, E and a data. Such plots are satisfactorily
linear for all of the adenine systems at potentials of
-0.5V or more negative. A set of q vs. θ plots is
shown in Figure 6 for dATP. Deviations from eq (16)
become quite pronounced for deoxyadenosine at potentials more positive than -0.6V. The fact that anodic
desorption peaks are not observed for any adenine species implies that at potentials more positive than
about -0.5V the Frumkin model does not hold. Hansen
(55) in a recent study of the adsorption of various
aromatic molecules has attributed the suppression of
anodic desorption peaks to specific anion adsorption.
The consistency of π, E, a results and the linearity
of q vs. θ plots at fixed potentials more negative
than -0.5V for most adenine systems makes it unlikely
that there are systematic errors in the measured capacitance data which render the derived π and q values
unreliable.

At pH 9 adenine and the adenine moiety in its
deoxynucleoside and deoxynucleotides is almost entirely in its neutral state (58).

The results of analysis of the capacitance curves
of adenine are shown in Table 1. These results reveal
that the attraction coefficient, α, is small and positive indicating a small lateral attractive interaction
between the adsorbed adenine molecules. The area occupied per adenine molecule on the electrode surface
is $55\pm4\text{Å}^2$. Adenine is a planar molecule as shown
from its crystal structure (59) and has an area of
about 42Å^2. However, considering the unsymmetrical
shape of this molecule and the resultant difficulty

Table 1. Parameters of the generalized Frumkin equation for adenine, deoxyadenosine and deoxyadenosine-5'-phosphates at pH 9[a]

Compound	α	$B_0 \times 10^{-3}$/ℓ/mole	$\Delta G°$/cal.[b]	E_N[c]	Γ_m/mole $cm^{-2} \times 10^{10}$	Area/$Å^2$ per molecule
Adenine	0.54±0.21	1.77±0.17	-4428	-0.560±0.1	3.04	55±4
Deoxyadenosine	-0.52±0.44	15.0±3.5	-5695	-0.506±0.021	3.04	55±4
Deoxyadenosine-5'-monophosphate	-0.10±0.19[e]	4.24±0.80[e]	-4949[e]	-0.527±0.005[e]	2.45	67±8[e]
Deoxyadenosine-5'-diphosphate	-0.55[d]	4.08±0.56	-4923	-0.472±0.015	2.22	75±4
Deoxyadenosine-5'-triphosphate	-1.42±0.43	10.8±1.3	-5498	-0.514±0.007	2.06	80±7

[a] Borate buffer (see Experimental)

[b] Standard free energy of adsorption at the electrocapillary maximum potential for the electrolyte, based on infinite dilution standard states for the adsorbate, both in solution (at unit molarity) and on the surface (at unit value of θ). $\Delta G° = -RT\ln B_0$, where B_0 is expressed in ℓ/mole units.

[c] Electrocapillary maximum potential where $\theta = 1$ for each compound.

[d] α value inferred by use of eq. 16.

[e] The parameters for dAMP were calculated using the sequential equal equilibrium constant model discussed in the text and a value for the equilibrium constant of 110 ℓ/mole.

in packing on the surface, we estimate that the actual area occupied by one adenine molecule should be
50-60Å². This implies therefore that in the first
adsorption region, adenine is adsorbed in a flat orientation on the electrode surface. Any electrode
surface not covered by adenine would be available to
water dipoles or other inorganic ions.

In the case of deoxyadenosine, the α value shifts
somewhat more negative compared to adenine (Table 1),
but the area occupied by one molecule is the same as
for adenine. This implies that the deoxyribose group
is tilted away from the electrode surface.

The attraction coefficient, α, shifts systemati-
cally more negative for the series dAMP, dADP, dATP.
Since the phosphate groups are extensively ionized at
pH 9, this effect is quite reasonable since a repul-
sive interaction between the adsorbed anionic species
is expected. The area occupied per molecule for each
adenine mononucleotide is very similar at ca. 70-80Å²
(Table 1). This area is only about 30 percent greater
than is observed for unsubstituted adenine or deoxyadenosine and hence indicates that the phosphate
groups are also largely directed away from the electrode surface.

In all of the systems except dAMP, the analysis
of π, E, a data was carried through assuming that concentrations of the organic solute could be substituted
for solute activity. However, in the case of dAMP,
considerable negative curvature occurs in plots of π
vs. log concentration at concentrations greater than
about 5mM (Figure 7). The best fit of these data with
the 5-parameter least squares program (at concentrations up to 20mM) leads to an anomalously large surface area for dAMP in the adsorbed monolayer (~ 126Å²).
A much improved correlation results if it is assumed
that dAMP partially associates according to the scheme:
monomer + monomer = dimer, monomer + dimer = trimer,
monomer + trimer = tetrameter, etc., and that the
activity of dAMP can be equated to the concentration
of monomer. For simplicity, the equilibrium constant
for each association step has been treated as a constant, K, (60). In fitting the dAMP data, K is considered to be a sixth adjustable parameter. The rms
deviation in π is reduced to nearly half its least
squares value from the 5-parameter optimization, and
a reasonable value of the area (67 ± 8Å²/molecule) is
obtained for dAMP when the least squares value of the
association constant (K = 110 ± 26ℓ/mole) is used to
calculate the activity of dAMP from its formal concentration. The solid curve in Figure 5C was calculated

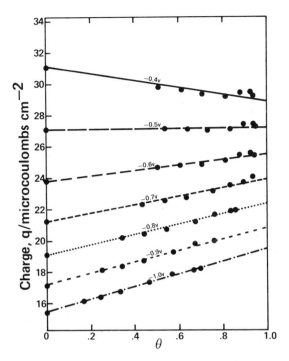

Figure 6. Surface charge vs θ plots for dATP at various potentials at the DME

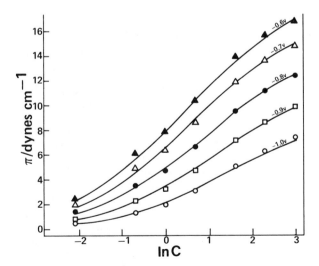

Figure 7. Surface spreading pressure π vs. ln concentration of dAMP

using K = 110ℓ/mole and using the five constants in
the generalized Frumkin equation reported in Table 1.
We have not been able to find any reports of association of dAMP. However, vapor pressure osmometry and
sedimentation equilibrium experiments indicate that
adenosine-5'-monophosphate (AMP) is highly associated
in the 0-50m$\underline{M}$ concentration range (61,62).
        The dependence of adsorption on potential is shown
in Figure 8 where it is clear that all compounds exhibit maximal adsorption at close to -0.5V. At more
positive or negative potentials adsorption becomes less
pronounced. These curves could have been extended to
about -1.5 to -1.6V where adsorption of the adenine
species no longer occurs to any appreciable extent.
        The free energy of adsorption at the E.C.M. for
the electrolyte ($\Delta G° = -RTlnB_o$, Table 1) exhibits no
very significant feature except an appreciable decrease in $\Delta G°$ between adenine and its deoxynucleoside
and monodeoxynucleotides. The $\Delta G°$ values simply indicate an increase in the surface activity of the
deoxynucleoside and deoxynucleotides over adenine.
        Using a capillary electrometer the interfacial
tension of the mercury-aqueous solution interface for
adenine and deoxyadenosine has been measured. Using
this technique, surface spreading pressure, π, may be
measured directly (eq 9). Some typical adsorption results obtained by electrocapillary measurements are
shown in Table 2. These results are in good agreement
with the values obtained by the more indirect capacitance method.
        Electrocapillary measurements have also been
used to study the adsorption of adenine and deoxyadenosine at concentrations where the capacitance pit is
observed (see Figures 3 and 4). It has not been possible yet to deduce the details of the adsorption processes but it is fairly easy to determine the limiting
slope of the π vs. ℓn a plots i.e., obtain Γ_{max} from
the Gibbs adsorption equation (eq 17). Typical results
are shown in Table 3. Calculations reveal that these
areas correspond closely to those expected for a vertical orientation of the adenine molecule at the electrode surface. It has been concluded therefore that
at concentrations where adenine or deoxyadenosine exhibit a capacitance pit there is a rearrangement of the
molecules, over a very sharply defined potential range,
from a flat arrangement on the surface to a perpendicular arrangement.
        In order to gain some additional insights into
the orientation of adenine molecules with respect to
the electrode surface when in the perpendicular orien-

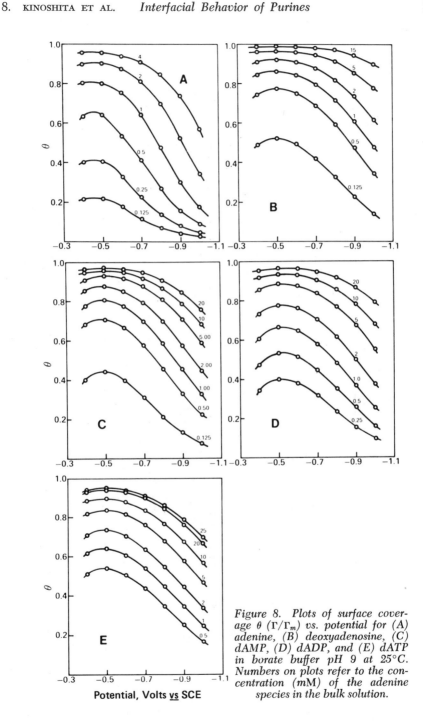

Figure 8.   Plots of surface coverage θ (Γ/Γ_m) vs. potential for (A) adenine, (B) deoxyadenosine, (C) dAMP, (D) dADP, and (E) dATP in borate buffer pH 9 at 25°C. Numbers on plots refer to the concentration (mM) of the adenine species in the bulk solution.

Potential, Volts vs SCE

Table 2. Parameters of the generalized Frumkin equation for adenine and deoxy-adenosine at pH 9 determined from electrocapillary measurements.[a]

Compound	α	B_o x10^{-3}/ℓ/mole	$\Delta G°$/ cal.	E_N/ V vs. SCE	Γ_m/mole cm^{-2}x10^{10}	Area/Å^2 per molecule
Adenine	0.49±0.12	1.26	-4226	-0.54	2.62	63±5
Deoxyadenosine	1.49±0.33	1.21	-4204	-0.50	3.09	54±3

[a]Terms are defined in Table 1.

Table 3. Γmax and area occupied per molecule at complete surface coverage for adenine and deoxyadenosine at potentials corresponding to the anomolous capacitance pit in borate buffer pH 9.

Compound	Maximum Potential Range/Volt vs. SCE	Γ_m/mole/ cm^{-2}x10^{10}	Area/Å^2 per molecule
Adenine	-0.35 to -0.9	7.2±0.9	23±3
Deoxyadenosine	-1.1 to -1.45	5.5±1.3	30±7

[a]See Figures 3 and 4 for appearance of capacitance pits.

tation, the effect of methylation of the amino group
of adenine has been studied. Capacitance $\underline{vs}$. potential curves for 6-dimethyladenine are shown in Figure
9. 6-Methyladenine exhibits very similar curves with
no capacitance pit. Quite clearly, substitution of
one or both amino hydrogens completely inhibits formation of the capacitance pits. Some typical results
for 6-methyl- and 6-dimethyladenine are shown in Table
4. The general behavior of these two compounds is
very similar to adenine although the area occupied per
molecule is a little smaller in the case of 6-methyl-
adenine. In view of the fact that only the purines
and pyrimidines found naturally in nucleic acids exhibit capacitance pits (41-44), it has been tentatively
concluded that the structural functionalities
associated with the hydrogen bonding of complementary
base pairs in nucleic acids (Figure 10) are responsible for binding to the electrode surface where molecules are in their perpendicular orientation, perhaps
as shown in Figure 11.

Ellipsometric Studies

Ellipsometry is based on the fact that when el-
liptically polarized light is incident on a film-
covered, highly reflective surface the polarized state
of the reflected light is different from that observed
in the case of reflection from a film-free surface.
The magnitude of the changes in the polarization state
induced by the surface film depends on the thickness
and the optical constants of the adsorbed film material. Some preliminary ellipsometric measurements have
been made on aqueous solutions of adenine and deoxyadenosine in contact with a plane (or nearly plane)
mercury electrode surface. A typical experimental arrangement used in these studies is shown in Figure 2.
Only qualitative or semi-quantitative studies are possible with the arrangement since difficulties are en-
countered with maintaining an absolutely flat mercury
surface and avoiding bubble formation at very negative
potentials. However, some of the potential capabilities of electroellipsometry can be illustrated by
giving the results of a simple experiment with deoxyadenosine. Figure 12 shows the capacitance $\underline{vs}$. potential curve for a 35m$\underline{M}$ solution of deoxyadenosine at
pH 7 at the DME. It will be recalled that electrocapillary and capacitance results suggest that between
about -0.2 to $\underline{ca}$. -1.1V deoxyadenosine is adsorbed
flat on the electrode, while in the capacitance pit

Table 4. Parameters of the generalized Frumkin equation for 6-methyladenine and 6-dimethyladenine at pH 9 determined by capacitance and electrocapillary measurements.

Compound	α	$B_o \times 10^{-3}$ / ℓ/mole	$\Delta G°$ / cal.	E_N / V vs. SCE	Γ_m/mole $cm^{-2} \times 10^{10}$	Area/$Å^2$ per molecule
6-Methyladenine	0.85±0.3 [b]	1.0±0.2 [b]	-4091 [b]	-0.549 [b]	3.83±0.24 [b]	43±3 [b]
	0.40±0.2 [c]	4.2±0.5 [c]	-4941 [c]	-0.549 [c]	3.63±0.16 [c]	46±2 [c]
6-Dimethyl-	0.5±0.6 [c]	14±0.5 [c]	-5695 [c]	-0.40 [c]	2.52±0.2 [c]	66±5 [c]

[a] For definition of terms see Table 1 and text.

[b] Determined by electrocapillary measurements.

[c] Determined by capacitance measurements.

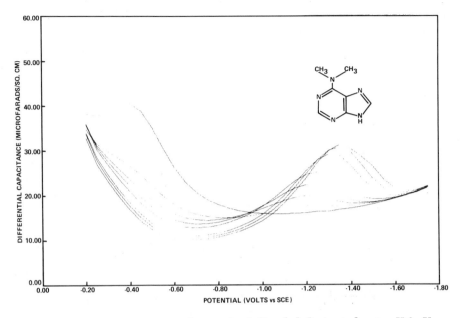

Figure 9.   Capacitance vs. potential curves for 6-dimethyladenine in borate pH 9. Upper trace is background electrolyte, lowest trace is 19.7mM 6-dimethyladenine.

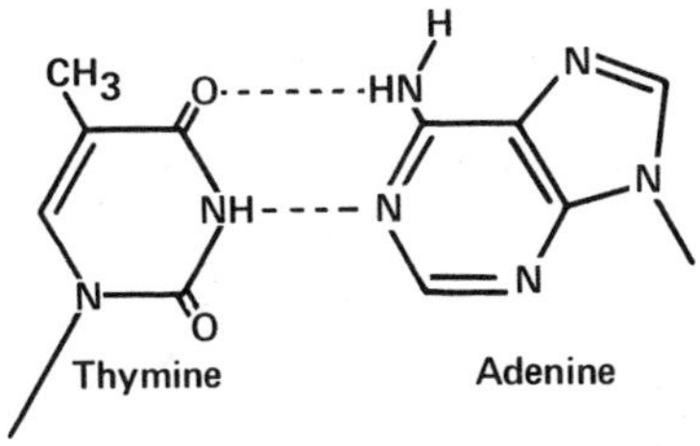

Figure 10.   Hydrogen bonding between thymine and adenine in DNA

Figure 11.   Possible modes of binding of adenine to the mercury electrode surface when in perpendicular orientation in region of anomalous capacitance pits

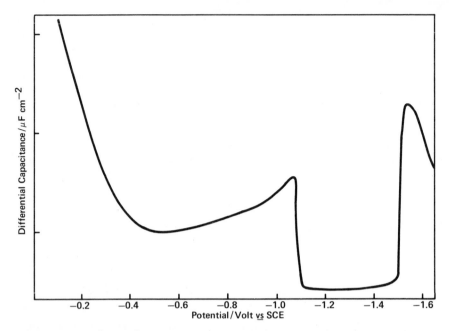

Figure 12. *Differential capacitance curve of 35mM deoxyadenosine in McIlvaine buffer pH 7*

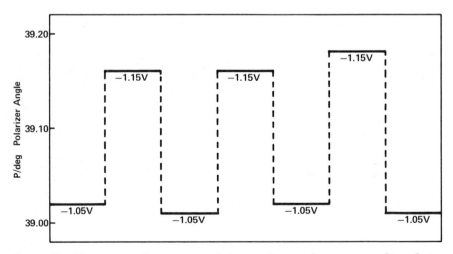

Figure 13. *Variation in ellipsometric polarizer angle at a plane mercury electrode in 35mM deoxyadenosine in McIlvaine buffer pH 7.0 as the potential is switched from −1.05V to −1.15V vs. SCE*

region (-1.1V to -1.5V) it is adsorbed perpendicular to the electrode surface. The variation in the ellipsometer polarizer angle (P) for the same solution at a plane mercury electrode as the electrode potential is varied between -1.05V (outside the capacitance pit) and -1.15V (inside the capacitance pit) is shown in Figure 13. P varies consistently by about 0.15° (absolute) as the potential is switched back and forth.

Assuming a reasonable refractive index of 1.48 to 1.50 for deoxyadenosine, this change in P corresponds to a change in thickness of about +5Å for the potential change from -1.05V to -1.15V (using the calculational methods of McCrackin [63]. Such an increment in film thickness agrees well with that inferred from capacitance and electrocapillary measurements, i.e., expected on the basis of a change from a flat (-1.05V) to perpendicular (-1.15V) arrangement.

Acknowledgements: The authors would like to acknowledge financial support of this work through National Institutes of Health Grant No. GM 21034. We would also like to thank Dr. Robert S. Hansen of the Iowa State University for his valuable comments and suggestions regarding some aspects of this work. The valuable assistance of Dr. Eric Enwall during some of the initial ellipsometric studies is also gratefully acknowledged.

Literature Cited
1. Hill, T. L., J. Amer Chem. Soc. (1958), 80, 2142.
2. O'Konski, C. T. and Katchalsky, A., Biophysical J. (1965), 5, 667.
3. Neumann, E., and Katchalsky, A., Proc. Nat. Acad. Sci., U.S.A. (1972), 69, 993.
4. Roberts, R. B. and Flexner, L. B., Quart Rev. Biophys. (1969), 2, 135.
5. Katchalsky, A., and Neumann, E., Intern. J. Neurosci. (1972), 3, 175.
6. Katchalsky, A. and Oplatka A., J. Med. Sci. (1966) 2, 4.
7. Katz, B. in "Biophysical Science-A Study Program," p. 466, J. L. Oncley (ed.), 1959.
8. Ganesan, A. T. and Lederberg, L., Biochem. Biophys. Res. Commun. (1965), 18, 824.
9. Ryter, A., Bact. Rev. (1968), 32, 39.
10. Comings, D. E., Ann. J. Human Genetics (1968), 20, 440.
11. O'Brian, R. L., Sanyal, A. B., and Stanton, R. H., Exp. Cell. Res. (1972), 76, 106.

12. Jacob, F., Brenner, S., and Cuzin, F., Cold Spr. Harb. Symp. Quant. Biol. (1963), 28, 329.
13. Becker, R. O., J. Bone Joint Surg. (1961), 43A, 643.
14. Becker, R. O. and Murray, D. G., Clin. Ortho. (1970), 75, 1969.
15. Becker, R. O., Nature (1973), 235, 109.
16. Becker, R. O. and Spadaro, J. A., Bull. N.Y. Acad. Med. (1972), 48, 627.
17. Cope, F. W., Bull. Math. Biophys. (1963), 25, 165.
18. Cope, F. W., Arch. Biochem. Biophys. (1963), 103, 352.
19. Cope, F. W., J. Chem. Phys. (1964), 40, 2653.
20. Cope, F. W., Bull. Math. Biophys. (1965), 27, 237.
21. Cope, F. W., Experientia. Suppl. (1971), 18, 223.
22. Cope, F. W., in "Oxidases and Related Redox Systems," T. E. King, H. S. Mason and M. Morrison (Eds.), Wiley, New York, 1965.
23. Cope, F. W., and Straub, K. D., Bull. Math. Biophys. (1969), 31, 761.
24. Cope, F. W., Adv. Biol. Med. Phys. (1970), 13, 1.
25. Mohilner, D. M. in "Electroanalytical Chemistry," A. J. Bard (Ed.), Vol. 1, p. 241, Dekker, New York, 1966.
26. Pilla, A. A., Bioelectrochem. Bioenerg. (1974), 1, 227.
27. Eisenberg, S., Ben-or, S., and Doljanski, F., Exptl. Cell. Res. (1962), 26, 451.
28. Mayhew, E. and Weiss, L., Exptl. Cell Res. (1968), 50, 441.
29. Mayhew, E., J. Gen. Physiol. (1966), 49, 717.
30. Brent, T. P. and Forrester, J. A., Nature (1967), 215, 92.
31. Purdon, L., Ambrose, E. J., and Klein, G., Nature (1958), 181, 1586.
32. Ambrose, E. J., Proc. 7th Canadian Cancer Res. Cont., p. 247, Pergamon, Toronto, 1967.
33. Weiss, L. and Hauscha, T. S., Int. J. Cancer (1970), 6, 270.
34. Field, A. V., Tytell, A. A., Lampson, P. G., and Hilleman, M. R., Proc. Nat. Acad. Sci. U.S.A. (1967), 58, 1004.
35. Schell, P. L., Biochem. Biophys. Acta (1971), 240, 472.
36. Janik, B. and Sommer, R. G., Biopolymers (1973) 12, 2803.
37. Janik, B. and Elving, P. J., J. Amer. Chem. Soc. (1970), 92, 235.

38. Janik. B. and Elving, P. J., Chem. Rev. (1968), 68, 295.
39. Dryhurst, G. and Elving, P. J., Talanta (1969), 16, 855.
40. Vetterl, V., Coll Czech. Chem Commun. (1966), 31, 2105.
41. Vetterl, V., Coll. Czech. Chem Commun. (1969), 34, 673.
42. Vetterl, V., J. Electroanal. Chem. (1968), 19, 169.
43. Vetterl, V., Biophysik (1968), 5, 255.
44. Lorenz, W., Z. Elektrochem (1958), 62, 192.
45. Webb, J. W., Janik, B., and Elving, P. J., J. Amer. Chem. Soc. (1973), 95, 991.
46. Krznarik, D., Valenta, P., and Nürnberg, H. W., J. Electroanal. Chem. (1975), 65, 863.
47. Kinoshita, H., Christian, S. D., and Dryhurst, G., J. Electroanal. Chem., submitted for publication (1976).
48. Damaskin, B. B., Petrii, A. A., and Batrakov, V., "Adsorption of Organic Compounds on Electrodes," p. 16, Plenum, New York, 1971.
49. Grahame, D. C., Coffin, E. M., Cummings, J. P., and Poth, M. A., J. Amer. Chem. Soc. (1952), 74, 1207.
50. Frumkin, A. N., Z. Phys. Chem. (1925), 116, 466.
51. Broadhead, D. E., Baikerikar, K. G., and Hansen, R. S., J. Phys. Chem. (1976), 80, 370.
52. Reference 48, p. 113.
53. Reference 48, p. 112.
54. Our computational program employs an optimizing program written by Dr. E. Enwall and incorporating an algorithm given by Marquardt, D. W., J. Soc. Indust. Appl. Math. (1963), II, 431.
55. Hansen, R. S., Personal communication.
56. Baikerikar, K. G., and Hansen, R. S., J. Coll. Interface Sci. (1975), 52, 277.
57. Reference 51, p. 70.
58. Alberty, R. A., Smith, R. M., and Bock, R. M., J. Biol. Chem. (1951), 193, 425.
59. Donohue, J., Arch. Biochem. Biophys. (1968), 128, 591.
60. Ts'o, P. O. P., in "Basic Principles in Nucleic Acid Chemistry," Vol. 1, pp. 537-543, Academic, New York, 1974.
61. Schweizer, M. P., Broom, A. D., Hollis, D. P. and Ts'o, P. O. P., J. Amer. Chem. Soc. (1968), 90, 1042.
62. Rossetti, G. P. and van Holde, K. E., Biochem. Biophys. Res. Commun. (1967), 26, 717.

63. McCrackin, F. L., "A Fortram Program for Analysis of Ellipsometric Measurements," NBS Technical Note No. 479, 1969.

Evaluation of Mediator-Titrants for the Indirect Coulometric Titration of Biocomponents

ROBERT SZENTRIMAY, PETER YEH, and THEODORE KUWANA

Department of Chemistry, Ohio State University, Columbus, Ohio 43210

In recent years we have been interested in the development and application of the indirect coulometric titration (ICT) method for the accurate assessment of stoichiometry ($\underline{n}$ value) and energetics ($E^{O'}$ value) of bioredox components such as cytochrome $\underline{c}$ oxidase and "blue" copper laccases. The initial reason for developing ICT was the advantage of being able to work with fairly small volumes under anaerobic conditions and to conveniently add electrochemical charge accurately and incrementally on nanoequivalent levels ($\underline{1}$). The optically transparent electrode also provided a means of easily acquiring spectral information during the titration ($\underline{2}$). Other advantages of ICT over classical potentiometric methods became obvious during our progress and these will be discussed subsequently.

The fundamental problem in the accurate assessment of $\underline{n}$ and $E^{O'}$ values of biocomponents is the slow heterogeneous electron transfer between the biocomponent and an indicator electrode such as platinum. This is particularly severe with large macromolecules where the redox site may be surrounded by some periphery structure such as a protein. Thus, one or more "mediators" are usually added to the solution so that redox coupling is enhanced between the biocomponent(s) and the electrode.

In the ICT method, a titrant (either a reductant or oxidant) is electrochemically generated to transfer charge to the biocomponent. For example, the reaction sequence for a reduction is:

Electrode reaction: $\quad M_{Ox} + ne^- = M_R \qquad E^{O'}{}_M \quad (1)$

Solution reaction: $\quad M_R + Enz_{Ox} = M_{Ox} + Enz_R \quad (2)$

where reaction (1) occurs at the electrode to generate the titrant, M_R which in turn reduces the biocomponent, Enz_{Ox}, to Enz_R. The equilibrium of reaction (2) is given by:

$$\Delta E^{O'} = E^{O'}_{Enz} - E^{O'}_{M} = RT \ln K_{eq} \qquad (3)$$

where $E^{O'}_{Enz}$ is the formal electrode potential for the half-reaction:

$$Enz_{Ox} + ne^- = Enz_R \qquad E^{O'}_{Enz} \qquad (4)$$

The $\underline{n}$ and $E^{O'}_{Enz}$ values are determined from the "best" fit between the experimental and computer simulated plots of the electrochemical charge, q, and the change in the optical absorbance, ΔA, of a biocomponent(s). As in potentiometry, one or more mediators may be added in small but known quantities to accelerate the attainment of equilibrium. Or alternatively, a titrant may be used whose $E^{O'}$ value is sufficiently close to that of the biocomponent (usually within 180 mv) so that it acts as a mediator. Thus, we have chosen to call redox titrants employed in the ICT method as mediator-titrants (M-T's).

The earlier choices of M-T's were those employed in potentiometric titrations of biocomponents, notably those reported for studies of components in the respiratory system, or those whose redox properties seemed suitable for M-T's as known from our experience or from the electrochemical literature. However, no systematic study has been previously reported to experimentally evaluate and compile a list of redox compounds which could serve as M-T's. Such a list would be particularly valuable to those examining bioredox components at potentials where suitable M-T's are presently unavailable or where well known mediators or M-T's have failed to give reproducible results for a particular biocomponent. Thus, a long-range objective of our laboratory has been to compile a list of possible M-T's whose potentials are graded in increments of some 20 to 40 mV's over a total potential range of ca. +1.00 to -1.00 volt versus NHE. Such a compilation is still incomplete. In this paper, the experimental assessment of several possible redox compounds as M-T's along with their effectiveness in the ICT of test biocomponents will be reported.

The "ideal" properties sought for M-T's are listed in Table I. The constraints imposed by these properties are so restrictive that very few compounds fulfill all of them. There are, fortunately, situations where some properties are less important than others. For example, let us assume that M_R in reaction (1) and (2) is chemically unstable and decomposes with a half-life of ca. 1 hour to another product. If $E^{O'}_{M}$ is much less than $E^{O'}_{Enz}$ (K_{eq} of reaction (2) is large) and the forward rate of reaction (2) is fast, then the concentration of M_R will be small such that any loss through decomposition will be minimal. The situation is somewhat different, however, when a compound is used as a mediator in a potentiometric titration. The mediator in this case acts as a redox buffer and is most effective near a 1:1 ratio of oxidized to reduced species. Instability of one species may then be detrimental depending on the rate of decomposition, properties of the decomposition product, and whether the product interacts with the biocomponent. As a matter of fact, the most difficult property to assess in Table I is whether any of the redox species inhibit or interact with any particular biocomponent. A priori judgement is difficult. Also, in the ICT method, two M-T's may or may not be compatible with each other. To assess the above properties and problems with M-T's the methods of cyclic voltammetry, controlled potential coulometry, thin layer spectroelectrochemistry, coulometric titrations of M-T versus M-T, and M-T titrations of biocomponents, particularly the test system of cytochrome c̲, have been employed.

Primary attention will be devoted to discussion of various ferrocene and bipyridylium compounds as M-T's. Although evaluation of all the properties of various M-T's is still in progress, a listing of M-T's has been compiled and is herein included since others may find one or more of these M-T's useful.

TABLE I

Properties of "Ideal" Mediator-titrants

1. well-defined n̲ value (n̲ = 1 preferrable for most cases)

2. known $E^{o'}$ value under experimental conditions

3. fast heterogeneous and homogeneous electron transfer

4. soluble in aqueous media at or near pH 7.0

5. stable redox species

6. good optical window

7. does not inhibit or interact with biocomponents

Results And Discussion

Bipyridylium Salts (Viologens). The negative value of the electrode potential for the bipyridylium salts (viologens) makes them attractive as reductants. They have been previously employed as one electron reducing mediator-titrants in the ICT of biocomponents (1-4). Others have utilized these salts for coulometric and kinetics studies (5-7) as well. Steckhan (8) has evaluated the spectroelectrochemical characteristics of several viologens, the halide salts of 1,1'-dibenzyl-4,4'-bipyridylium, 1,1'-ethylene-2,2'-bipyridylium, 1,1'-dimethyl-4,4'-bipyridylium, and 1,1'-propylene-2,2'-bipyridylium dications using optically transparent electrodes (OTE's). The discussion to follow extends and illustrates the further applications of these viologens for use as M-T's in ICT of biocomponents at the nano-equivalent (10-100) levels. The properties of viologens examined to date are listed in Table II. The $E^{O'}$ values (pH = 7.0, phosphate buffer with ionic strength of 0.15) range from -358 mV to -556 mV for the first electron reduction (reaction (5)). A second reduction step (reaction (6)) reduces the radical cation to the neutral species. In most of these viologens, the neutral species,

$$V^{++} + e^- = V^{\ddot{+}} \qquad E^{O'}_1 \qquad \qquad (5)$$

$$V^{\ddot{+}} + e^- = V^O \qquad E^{O'}_2 \qquad \qquad (6)$$

V^O, is insoluble and is strongly adsorbed on the electrode surface. The radical cation, $V^{\overset{+}{\cdot}}$, also may form sparingly soluble salts, depending on the particular viologen and the counter ion present in solution. A typical current-potential (i-E) pattern for cyclic voltammetry at a tin oxide OTE at pH 7.0 is shown in Figure 1. The reverse oxidative wave for the neutral species varies and often shows typical characteristics for the electrolysis of adsorbed reactant. The 1st wave for the reduction of V^{++} to $V^{\ddot{+}}$ and the oxidation of the radical back to the dication is usually reversible. For the quantitative generation of the radical cation as a reductant for a solution reaction during coulometric titrations, it is imperative that the solution conditions and the potential be selected so that insolubility or adsorption does not occur. The separation between $E^{O'}_1$ and $E^{O'}_2$ is usually sufficient for most viologens that the radical cation can be quantitatively generated without interference from the neutral species. This has been demonstrated previously for methyl viologen by setting the potential of the chromoamperometric experiment at no more than 20 mv negative of $E^{O'}_1$ (8,9).

TABLE II

Electrochemical and Optical Properties of Viologens

Viologen Dications	$E_1^{o'}$ (mV vs. NHE)	$E_2^{o'}$ (mV vs. NHE)	λ_{max} (nm)	ϵ (cm^{-1} M^{-1})
1,1'-dibenzyl-4,4'-bipyridylium	-358[a]	-538[a]	595[a]	10,100[a]
1,1'-ethylene-2,2'-bipyridylium	-361[a]	-693[c]		
1,1'-bis(hydroxyethyl)-4,4'-bipyridylium	-408[b]	-753[b]	598[b]	9,300[b]
1,1'-dimethyl-4,4'-bipyridylium	-449[a]	-772[a]	605[a]	12,400[a]
1,1'-propylene-2,2'-bipyridylium	-556[a]	-816[c]	510[a]	14,900[a]

[a]from ref (8)

[b]this work

[c]from ref (43)

The initial steps in the preparation for ICT of a biocomponent are to quantitate the coulometric generation of the M-T(s) and to check on the anaerobicity of the electrolysis cell. Figure 2 shows the plot of the change in the optical absorbance, ΔA, at 595 nm versus the electrochemical charge, q, for the generation of BV^+ at a tin oxide OTE. The potential applied to the OTE was -0.60 V vs Ag/AgCl reference electrode. The subsequent removal of benzyl-viologen radical cation, $BV^+_\cdot$, as indicated by the decrease of ΔA in Figure 2 was accomplished through the electrolytic generation of molecular oxygen (3) at a platinum microelectrode (applied potential of +1.40 V vs Ag/AgCl reference). In the particular cell used, two to three minutes of solution stirring was required between each charge injection so that redox equilibrium could be attained throughout the solution. A spectrum was obtained after each equilibration and before the next charge injection. For reductions, a charge correction of 15 $\pm$ 3% was required to correct for background charge which was evaluated from coulometric experiments in the absence of the M-T. With some tin oxide OTE's, this correction would be as low as 2% which depended on the solution conditions for the experiment. It is therefore necessary to carefully evaluate this background correction for each run. With the 15% correction, the average n values calculated from the slopes (3,4) of the ΔA-q plots were 1.08 $\pm$ 0.03 and 1.08 $\pm$ 0.06 for oxidation and reduction of BV, respectively. The background correction for the Pt microelectrode was less than 1% of the total charge for O_2 generation.

It is also advisable to titrate a M-T against another one. Such a coulometric titration is illustrated in Figure 3 where the optical absorbance of the 1,1'-bis(hydroxymethyl) ferricinium ion ($BHMF^+$) at wavelength of 640 nm is being followed. The $BHMF^+$ is quantitatively generated at a tin oxide OTE by applying +0.60 V vs Ag/AgCl reference electrode. The reduction of the ion was through the electrogeneration of $BV^+_\cdot$. Since larger increments of charge were employed for this titration as compared to the amount for most biocomponents, the background corrections are considerably less (2 $\pm$ 1%). The redox cycling of $BHMF^+/BHMF$ can be reproducibly repeated several times without any noticeable deviation in the ΔA-q curves. It should be noted that the sequence of the $BHMF^+$ reaction with $BV^+_\cdot$ is typical of the ec catalytic regeneration mechanism:

$$BV^{++} + e^- = BV^+_\cdot \tag{7}$$

$$BHMF^+ + BV^+_\cdot = BV^{++} + BHMF$$

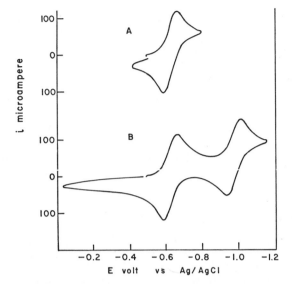

Figure 1. Typical cyclic voltammetric i–E curve for viologens. Reduction of .96mM concentration of 1,1′-dihydroxyethyl-4,4′-bipyridylium chloride in phosphate buffer pH 7.0 (ionic strength 0.15) at tin oxide OTE; scan rate 96 mv/S. (A) i–E of 1st wave only; (B) i–E of 1st and 2nd reduction wave.

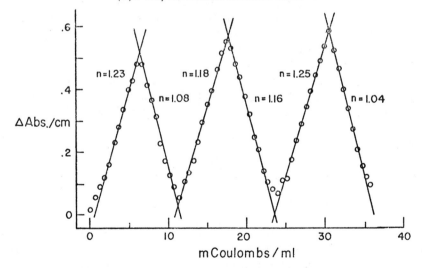

Figure 2. Change in optical absorbance vs. charge plot for the generation and removal of benzylviologen radical cation. Concentration of benzylviologen chloride 1.00mM in phosphate buffer at pH 7.0 (ionic strength 0.15); monitoring wavelength of 595 nm using a cell with an optical path length of 1.25 cm, cell volume 1.33 ml. Increase in absorbance corresponds to the generation of the benzylviologen radical cation.

Potentiometric measurements can be made concurrently during coulometry of BV^+ or $BHMF^+$ generation by monitoring the electrode potential of a second small Pt electrode in the cell. The $E^{O'}$ value of the $BHMF^+/BHMF$ couple determined in this manner will be discussed shortly.

In Figure 4, typical results are presented for the ICT titration of the test biocomponent, cytochrome c (cyto c). The cyto c concentration was 21.2 μM and the cell volume was 1.33 ml (4). The redox concentration of cyto c was monitored by following the optical absorbance at a wavelength of 550 nm after each incremental addition of charge. The two M-T's discussed above were employed for this ICT redox cycling. Sequential reductive and oxidative cycling could be repeated as many as eight times without any obvious change to the shape of the ΔA-q curves. Each charge increment in this titration was ca. 4.14 nanoequivalents. The average n values for four consecutive cycles as shown in Figure 4 were 1.01 ± 0.01 (6% corrected for background charge) and 1.06 ± 0.04 (3% corrected for background charge) for reductions and oxidations, respectively. The quantitation for the ICT of cyto c using these two M-T's is excellent and is in good agreement with previous results (1). The abrupt change in the slope of the ΔA-q plot during reduction indicates that cyto c is fully reduced and that an excess of BV^+ had been generated. This change conveniently marks the end point for the titration. Thus, in the oxidations, the excess must be removed prior to the commencement of the oxidation of reduced cyto c.

When an excess of BV^+ was present in the solution, a slow loss of the radical was found as evidenced by the decrease of the optical absorbance. The rate of loss was about 0.002 to 0.004 absorbance unit per minute or corresponded to about 0.2 to 0.4 nanomoles per minute. This loss may, in part, explain the 8% error in the coulometry of BV^+ found earlier (results as shown in Figure 2). However, to date we have found the precision in the ICT of biocomponents to be within 5% using this M-T, as expected since BV^+ is being consumed as generated. The reason for the loss of this radical when it is in excess is presently unexplained.

Both methyl and benzyl viologen have been extensively used as reductive M-T's in the titration of the heme proteins, cyto c and cytochrome c oxidase, with good results. Preliminary results from the ICT of tree laccase (10) and of heme proteins in submitochrondrial particles have produced well defined ΔA-q curves which are interpetable and assignable to the expected components. At present, there is no a-priori reason why the other viologens listed in Table II will not act as satisfactory M-T's for

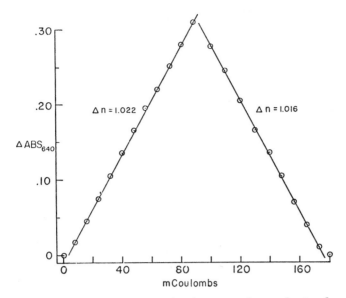

Figure 3. Change in optical absorbance vs. charge plot for the generation and removal of 1,1'-bis(hydroxymethyl)ferricinium ion. Concentration of 1,1'-bis(hydroxymethyl)ferrocene 1.02mM and 1.22mM benzylviologen chloride in phosphate buffer at pH 7.0; cell parameters same as those used in experiment shown in Figure 2. Increase in absorbance corresponds to the generation of the ferricinium ion.

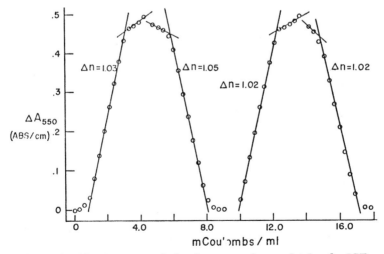

Figure 4. Change in optical absorbance vs. charge plot for the ICT of 21.2 μM cytochrome c. The M-T's are those used for the experiment shown in Figure 3 using same experimental conditions. Increase in absorbance at 550 nm corresponds to the reduction of cytochrome c.

TABLE III

Electrochemical and Optical Properties of Ferrocenes

Ferrocene derivatives	E°'(mV vs. NHE)	ΔE_p(mV)	Solubility[a]	λ_{max}(nm)[b]	ϵ(cm⁻¹ M⁻¹)[b]	$t_{1/2}$(hr)[b]
1,1'-dimethyl	341±9	51	I,D	655		32
acetic acid	365±10[c]	57	S	630	370	4.3
hydroxyethyl	402±10	58	S	625		>25
ferrocene (parent)	422±6[d]	62[d]	I,D	615[d]	335[d]	14[d]
1,1'-bis(hydroxymethyl)	465±5[c]	57	S	638	385	≥24
hydroxy-2-phenylethyl	480±5	61	I,D			
monocarboxylic acid	530±10[c]	64	S	630	420	0.50
chloro	589±8	57	I,D			
ferrocenylmethyl tri-methyl ammonium salt[e]	627±11	57	S	625		1
1,1'-dicarboxylic acid	644±12	40	S			

aThe symbol S indicates water solubility; I,D means insoluble and detergent solubilized using 3% Tween-20.

[b] The wavelength maximum for the oxidized form, the ferriciniums.

[c] OTTLE = Optically Transparent Thin Layer Electrode.

[d] Taken from Ref. (17).

[e] The counter anion was perchlorate.

reductive titrations.

Ferrocene and Ferrocene Derivatives

The unique structure and properties of ferrocene and its derivatives have resulted in a great deal of theoretical and experimental studies during the last two decades. Of particular interest to us was the wide range of positive redox potentials accessible through the variety of substituents of ferrocene, the facile electron-transfer properties, the clear optical window in the visible region for ferrocenes in their reduced form, and the well defined $\underline{n}$ value of unity. Their formal redox potentials and polarographic $E_{1/2}$ values have been reported in various literature compilations ($\underline{11}$, $\underline{12}$, $\underline{13}$, $\underline{14}$). However, there are two major problems with the use of ferrocenes as M-T's. These are the limited solubility of some ferrocenes in their reduced form and the instability of the oxidized form, the ferricinium ions, in aqueous solutions, particularly near physiological pH's. The number of oxidants for biological applications have been quite limited with the most familiar ones being the metal cyanides (Fe, Mo, W) which can be deleterious with certain biocomponents at higher concentrations ($\underline{15}$, $\underline{16}$).

There have been few recent studies which encouraged the further examination of ferrocenes as M-T's. The limited solubility could be circumvented by solubilization in micelles as formed by non-ionic detergent such as Tween-20 ($\underline{17}$). Using such solubilization, Fujihira, et al., demonstrated the ICT of reduced cyto $\underline{c}$ oxidase by electrogenerated ferricinium ion ($\underline{16}$).

The $E^{0'}$ values for various ferrocenes in phosphate buffer at pH of 7.0 are given in Table III. The ferrocenes in the Table span a range of potentials from +340 to +644 mV and were selected from a list of 24 ferrocenes which have been examined. Cyclic voltammetry at a Pt OTE was used for determination of the $E^{0'}$ values except as noted otherwise in the table. The trend in the $E^{0'}$ values is in agreement with that expected on the basis of substituent effects ($\underline{11}$, $\underline{12}$, $\underline{13}$, $\underline{14}$). The lower relative potentials for ferrocene monocarboxylic acid (FMCA) and ferrocene acetic acid (FAA) may be due to the acid-base equilibrium in which the basic form predominates at pH 7.0. Thus, these two compounds are more easily oxidized than expected from substituent effect considerations alone and are in agreement with calculations of Penden, et al. ($\underline{18}$). The insoluble compounds (labelled I, D) in Table III were solubilized using 3% Tween 20 according to the procedure of Yeh and Kuwana ($\underline{17}$). The cyclic voltammetric i-E curves of these detergent solubilized ferrocenes exhibited rever-

sible to nearly reversible behavior. The difference between the
oxidative and reductive peak potentials (ΔE_p) was less than 70 mV
for these ferrocenes.

Three ferrocenes, FMCA, FAA and BHMF (1,1'-bis-(hydroxy-
methy'.)ferrocene) with $E^{O'}$ values of +530, 365 and 465, respectively were examined quite thoroughly, primarily because of their
solubility (up to about 10 m$\underline{M}$ in phosphate buffer at pH 7.0) and
their attractive redox potentials.

Typical cyclic i-E curves for BHMF at both tin oxide OTE and
Pt electrodes are shown for comparison purposes in Figure 5. The
"irreversible" i-E for the tin oxide electrode is quite typical for
most ferrocenes at this electrode. The reversibility varies with
each electrode and the pH ($\underline{19}$). Thus, the potential required for
the diffusion controlled oxidations of ferrocenes must be prede-
termined for each coulometric experiment.

The stability of these three ferricinium ions, $FMCA^+$, FAA^+
and $BHMF^+$ was determined by monitoring their optical absorbance
after they were generated by coulometry. In Figure 6 the plots of
absorbance, $\underline{A}$, versus time are shown for these ions in aqueous
solution at pH 7.0. The wavelength was set at 630 or 640 nm
which is the long wavelength maximum characteristic of the
ferriciniums (see Figure 5 for spectra of BHMF and $BHMF^+$). These
plots are characteristic of first order kinetics and the half-lives
are 0.50, 4.3 and $\geq$ 24 hours for $FMCA^+$, FAA^+ and $BHMF^+$, respectively. It is interesting that, when all of the ions have been
completely lost, 50-75% of the initial concentration can be regenerated again by oxidative electrolysis. These results tend to
support the interpretation by Penden, et al. ($\underline{20}$) that the ferricinium ion undergoes a hydrolysis reaction involving a disproportionation mechanism. This disproportionation produces one-third
ferric hydroxide and the remainder, the parent ferrocene. The
half-lives for other ferriciniums are listed in Table III. Irrespective of the instability of ferricinium ions, they can be employed
as oxidative M-T's if their $E^{O'}$ values are sufficiently positive
of the biocomponent so that their equilibrium concentration remains relatively small during the coulometric titrations. Potentiometric, voltammetric and spectra data for BHMF, FMCA and FAA
(concentrations $\underline{ca}$. 10 m$\underline{M}$) were also obtained using the spectroelectrochemical method at the transparent thin layer cell using
a gold minigrid electrode. The data are summarized in Table III.
The thin layer experimental procedures as described by Heineman
($\underline{21}$) were adopted.

The effect of the differing $E^{O'}$ values of these three ferrocenes
is clearly illustrated in Figure 7 which shows the $\Delta\underline{A}$-q plots for
the ICT of cyto $\underline{c}$ ($\underline{ca}$. 20 $\mu\underline{M}$). The $\Delta\underline{A}$-q curve for each titration

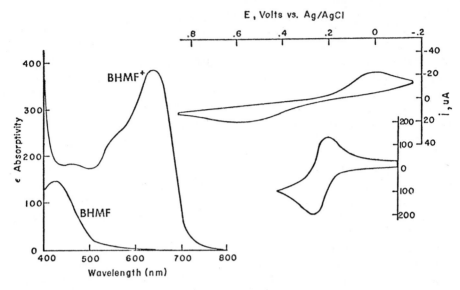

Figure 5.   (A) (left) Absorption spectra of the 1,1'-bis(hydroxymethyl)ferrocene and its oxidized form, the ferricinium ion. (B) (right) Cyclic voltammetric i–E curves for 1.0mM 1,1'-bis(hydroxymethyl)ferrocene at tin oxide OTE (top of figure). Area of electrode 0.8 cm^2 (bottom of figure is for same compounds at Pt electrode, area of electrode ca. 2 cm^2). Solution contains phosphate buffer at pH 7.0.

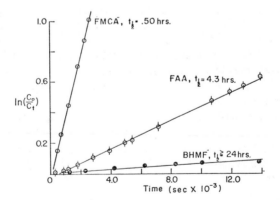

Figure 6.   First order kinetic plots for the loss of the ferricinium ions. FMCA$^+$ = O; FAA$^+$ = Ọ; BHMF$^+$ = ●. Concentrations were 1–2mM in phosphate buffer at pH 7.0.

has been normalized to a common end-point (100% ΔA) and charge scale (equivalents/mole). The data points at ΔA greater than 100% represent the generation and subsequent removal of excess BV^{+} ($E^{O'}$ = -358 mV vs NHE) which was the electrogenerated reductant in all three cases. With $FMCA^{+}$ as the oxidant ($E^{O'}$ = +530 mV vs NHE), the ΔA-q plot is nearly linear since its $E^{O'}$ is 275 mV more positive than that of cyto c. With $BHMF^{+}$, the ΔA-q plot shows slight curvature as cyto c becomes nearly fully oxidized. This curvature reflects the equilibrium between $BHMF^{+}/BHMF$ ($E^{O'}$ = 465 mV vs NHE) and cyto c_{Ox}/cyto c_{R} since the K_{eq} (K_{eq} = $10^{3.56}$) is relatively small. The situation becomes more accentuated with FAA^{+}/FAA ($E^{O'}$ = 365 mV vs NHE) since the K_{eq} is only $10^{1.86}$. There is pronounced curvature along the entire ΔA-q plot. The solid lines through the data points in Figure 7 were computer calculated assuming the $E^{O'}$ of cyto c to be 255 mV vs. NHE. The good agreement between experimental and calculated ΔA-q plots give confidence of the attainment of redox equilibrium and that these ferrocenes do not inhibit or interact chemically with cyto c.

Similar data are presented in Figure 8 for the quantitative ICT of cytochrome c oxidase using BV^{+} as reductant and $BHMF^{+}$ as oxidant. The solid lines in this figure are the computer simulated ΔA-q curves which were calculated by assuming the $E^{O'}$ values of 350 mV vs NHE (high potential copper-heme pair) and 210 mV vs NHE (low potential copper-heme pair) (3).

The equal contribution of each heme to the total absorbance change at the monitoring wavelength of 604 nm was assumed. Comparison between the reductive and oxidative ΔA-q curves show a small degree of hysteresis which is indicative of some irreversibility. Reversibility had been previously reported for this enzyme using detergent solubilized ferrocene which was electrolyzed to ferricinium ion as an oxidant (16). However, both $FMCA^{+}$ and $BHMF^{+}$ oxidations of cyto c oxidase have shown behavior varying between near reversibility to the type of anisotropy shown in Figure 8. The exact nature of this anisotropy is not known. Numerous interpretations of the redox behavior of cyto c oxidase have been proposed in the literature including various states of oxidized and oxygenated oxidase (22). Schroedel and Hartzell (23) have recently interpreted these types of titration curves to a redox mechanism which rationalizes difference between the reductive and oxidative curves.

The electrogeneration of ferricinium ions or the chemical oxidation of ferrocenes to ferriciniums that are fairly stable provides oxidants which possess many of the properties of ideal M-T's. They will greatly expand the arsenal of oxidants which were previously limited to few metal cyanides and metal complexes

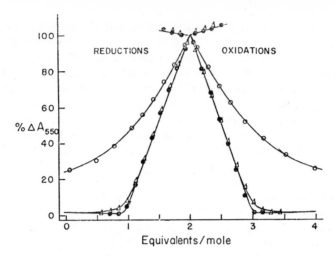

Figure 7. Normalized change in absorbance vs. charge plots for the ICT of cytochrome c using benzylviologen and different ferriciniums. ○, 25μM cytochrome c and 0.60 mM FAA; △, 20μM cytochrome c and 1.02 mM BHMF; ●, 21μM cytochrome c and 0.75 mM FMCA. 1–2 mM benzylviologen using phosphate buffer at pH 7.0 (ionic strength 0.15). Solid lines drawn through the experimental points are computer simulated curves for the experiment.

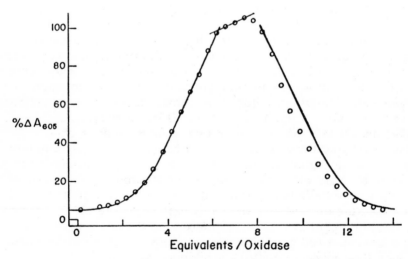

Figure 8. Normalized change in optical absorbance vs. charge plot for the ICT of cytochrome c oxidase. 15.6μM cytochrome c oxidase (100% ΔA = .375 a.u./cm); 0.33mM 1,1'-bis(hydroxymethyl)ferrocene and 1.0mM benzylviologen; phosphate buffer at pH 7.0 (ionic strength 0.15); cell parameters same as those in experiment shown in Figure 2. Solid lines are computer simulated curves assuming the oxidase to be n = 4 (see text for E°′ values).

such as those derived from bipyridyl ligands. Preliminary experiments with the oxidation of tree laccase and the heme proteins in the submitochondrial particles using ferriciniums have produced results interpretable in view of previous works (10).

Molybdenum Octacyanide

In search of M-T's with $E^{O'}$ values much more positive than the commonly used ferricyanide, molybdenum octacyanide $(Mo(CN)_8^{-3}/Mo(CN)_8^{-4})$ was evaluated. It had been utilized previously for potentiometric and kinetic studies of the laccases (24, 25). The generation of $Mo(CN)_8^{-3}$ as a catalytic oxidant at micromolar levels had already been reported using various electrodes (26, 27, 28). The commonly used tin oxide OTE in our laboratory was thought to be superior to other electrodes for oxidations at these potentials since tin oxide has a high overpotential and has a surface that is already oxidized. Laitinen and Conley (29) have reported the quantitative generation of Ag(II) with the current efficiencies being higher at this electrode compared to either Pt or Au. Experimental care must be taken in handling molybdenum octacyanide due to its photosensitivity (30).

A typical cyclic i-E curve for 10.6 m$\underline{M}$ $Mo(CN)_8^{-4}$ using a tin oxide thin layer spectroelectrochemical cell is shown in Figure 9. The shape of the i-E curve is characteristic of a reversible thin layer electrochemical system with uncompensated iR drop.* Spectra obtained concurrently during potential-step electrolysis of the $Mo(CN)_8^{-4}$, according to the method of Heineman, et al. (21) are also presented in Figure 9. If a nernstian plot of the applied potential (E_{appl}) versus the logarithmic ratio of the oxidized to reduced forms of the molybdenum octacyanide as determined by the change in the optical absorbance at 388 nm, are plotted, a linear line results. The average $E^{O'}$ determined from four such experiments gave a value of $+798 \pm 3$ mV vs. NHE. This value compares well with those previously reported at the same ionic strength (30, 31). The $\underline{n}$ value calculated from the slope of the plots is $1.00\pm.02$. Thus, the molybdenum octacyanide appeared as a good oxidizing M-T (See Figure 10).

---

*This is confirmed by the fact that $E^{O'}$, calculated as $E^{O'} = E_{pa} + E_{pc}/2$, was found to be independent of scan rate. Here E_{pa} and E_{pc} are the anodic and cathodic peak potentials for the cyclic voltammograms.

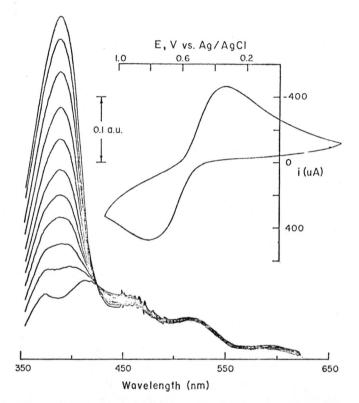

*Figure 9.   Thin layer spectroelectrochemical data for molybde-
num octacyanide. (upper right) Cyclic voltammetric i–E curve
for $Mo(CN)_8^{-3/-4}$ in a tin oxide thin layer spectroelectrochemical
cell. (left) Spectra obtained during incremental addition of
charge for oxidation of $Mo(CN)_8^{-4}$; 10.63mM molybdenum octacyanide in 0.50M NaCl, phosphate buffer at pH 7.0. The wavy
baseline on the spectrum is from the tin oxide electrode (interference pattern).*

Our initial test of molybdenum octacyanide in a M-T versus M-T coulometric titration failed when BV^{++} was used as the other M-T because precipitation occurred as BV^{+} was electrogenerated. After examining various other possible reductants, the electro-reduction of anthraquinone-2-sulfonate at tin oxide OTE proved to be compatible with the molybdenum octacyanide. Using the above M-T's, the ICT quantitation was accomplished for cyto c (Figure 11). The results from four reductive-oxidative cycles gave an average n value of 1.07 ± 0.04 (uncorrected, about 4% background contribution). The second ΔA-q curve shown in Figure 11 is for the reductive and oxidative titration of cyto c oxidase. The shape of this ΔA-q curve is considerably different from that obtained for oxidase titrated by BV^{+} and $BHMF^{+}$ (see Figure 8). It is evidence of either finite complexation by dissociated cyanide ion or by $Mo(CN)_8^{-4}$ of cyto c oxidase. In addition, the reduced enzyme is only partially active toward oxygen. Thus, there appears to be serious inhibition produced by the presence of the molybdenum octacyanide to cytochrome c oxidase. Irrespective of the previous results for the use of molybdenum octacyanide as an oxidant for the chemical titration of laccases (24), the absence of inhibition or interaction by this compound to the redox states of laccase still remains to be proven. Part of the success in using the molybdenum octacyanide in the laccase titrations may be due to the lower concentrations (less than 40 μM) employed. However, these ICT results should be indicative of the precautions which must be taken in the use of M-T's for biocomponents, particularly those that have complexing ligands which can be slowly and irreversibly dissociated and then taken up by the biocomponent(s).

The molybdenum octacyanide titration of cyto c oxidase serves to emphasize the importance of performing multiple ICT's using several M-T's for confirmation of n and $E^{O'}$ results.

Other M-T's

Our search and characterization of possible M-T's are still far from being completed. Some 60 redox compounds have been identified now as possible M-T's and characterization of these and many more is anticipated for future work. Table IV lists several compounds which have been reported and used as mediators or those which have been given preliminary screening in our laboratory and may prove to be useful. For example, the 2,2'-bipyridyl complexes of ruthenium, iron and osmium were examined because of their very positive formal potentials. However, the reduced forms of these metal complexes are highly colored in the visible region of the spectrum (see Table IV for λ_{max} and ϵ data)

TABLE IV

Redox Compounds Useful as Mediator-titrants

E°'(mV, NHE)[a,b]	Compound	n	ΔE$_p$(mV)[b]	Reduced Form		Oxidized Form	
				λ$_{max}$(nm)[a]	ε(cm⁻¹ M⁻¹)[a]	λ$_{max}$(nm)[a]	ε(cm⁻¹ M⁻¹)[a]
+1,272 (1,274)[34]	Ruthenium tris-(2,2'-bipyridine)	1	60	450 (445)[35]	16,000 (14,600)[35]	(418)[35]	(8,100)[35]
1,107 (1,120)[36]	Iron tris-(1,10-phenanthroline)	1	55	507 (510)[37]	11,000 (11,100)[37]	(590)[38]	(600)[38]
1,074 (1,096)[36]	Iron tris-(2,2'-bipyridine)	1	58	518 (522)[39]	8,800 (8,650)[39]	(610)[40]	(330)[40]
844 (877)[41]	Osmium tris-(2,2'-bipyridine)	1	61	475 (447)[41]	13,800 (13,700)[42]		
798 (774)[43]	Molybdenum octa-cyanide	1				388	1,365
418 (415)[44]	Iron hexacyanide	1	55			418	1,140
371 (366)[45]	N,N-Dimethyl-p-phenylenediamine	2	63			550 (550)[54]	350

TABLE IV (CONTINUED)

Redox Compounds Useful as Mediator-titrants

E°'(mV, NHE)[a,b]	Compound	n	ΔE_p(mV)[b]	Reduced Form		Oxidized Form	
				λ_{max}(nm)[a]	ϵ(cm^{-1} M^{-1})[a]	λ_{max}(nm)[a]	ϵ(cm^{-1} M^{-1})[a]
270 (270)[45]	N,N,N',N'-Tetramethyl-p-phenylenediamine	2	60			560 (565)[46]	12,000 (12,470)[46]
257 (240)[47]	2,3,4,5-Tetramethyl-p-phenylenediamine	2	40			473 (480)[54]	370
(227)[21]	2,6-Dichlorophenol-indophenol	2				(600)[21]	(2,060)[21]
157	1,2-Naphthoquinone	2	100			405	2,700
92 (80)[48]	Phenazine metho-sulphate	2	80			430	6,860
-3 (+33)[49]	5-Hydroxy-1,4-naphthoquinone	2	85			420	2,400
-10 (-53)[50]	Pyocyanine	2	70			682	4,090

TABLE IV (CONTINUED)

Redox Compounds Useful as Mediator-titrants

$E^{\circ\prime}$(mV, NHE)[a,b]	Compound	$\underline{n}$	ΔE_p(mV)[b]	Reduced Form λ_{max}(nm)[a]	ϵ(cm^{-1} M^{-1})[a]	Oxidized Form λ_{max}(nm)[a]	ϵ(cm^{-1} M^{-1})[a]
(−60)[51,c]	2-Amino-6,7-dihydro-4-pteridone	2					
−133 (−137)[52]	2-Amino-1,4-naphthoquinone	2	135			450	1,500
(−225)[58]	Anthraquinone-2-sulphonate	2					
(−660)[51,c]	2-Amino-4-pteridone	2					

[a] literature values are in parentheses.

[b] supporting electrolyte used is phosphate buffer (pH 7.0, ionic strength 0.15).

[c] ref. (51) gave $E_{1/2}$ values of −0.30 and −0.90 volt with respect to SCE for 2-amino-6,7-dihydro-4-pteridone and 2-amino-4-pteridone, respectively, in pH 9 borate buffer.

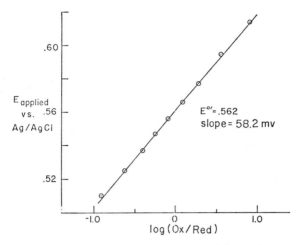

Figure 10. Plot of the applied potential ($E_{applied}$) vs. the optically determined logarithm of concentration ratio of oxidized to reduced molybdenum octacyanide. Experimental conditions same as those in Figure 9.

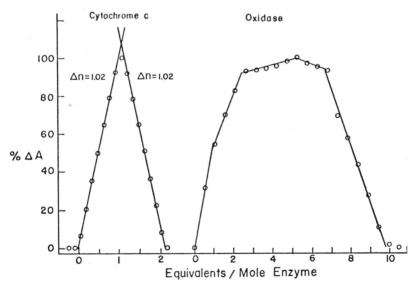

Figure 11. Normalized plots of optical absorbance vs. charge for the ICT of cytochrome c and cytochrome c oxidase with molybdenum octacyanide and anthraquinone 2-sulfonic acid. (left) ICT of cytochrome c ($22\mu M$); (right) ICT of cytochrome c oxidase ($12\mu M$). M-T's of 1.0mM $Mo(CN)_8^{-3}$ and 1.3mM anthraquinone 2-sulfonic acid. $13\mu M$ phenazine methosulfate added to the oxidase solution to insure equilibrium.

and may cause problems for optical monitoring of biocomponents.
The oxidized forms of these complexes may be quantitatively
generated at the tin oxide OTE and M-T vs M-T ICT's have been
performed using benzyl viologen as the other M-T with satisfactory results. However, oxidative ICT's of reduced cyto c with
these M-T's exhibited drawn-out ΔA-q curves indicative of oxidation of more than the iron heme moiety. It is interesting to note
that the previously discussed molybdenum cyanide gave quantitative n = 1 titration of cyto c which suggests that oxidation of
other parts of cyto c occurs at potentials above ca. +800 mV.

The substituted p-phenylenediamines have been previously
employed as mediators and Mackey (4) demonstrated the quantitative electrogeneration of the diimine in the case of tetra-methyl-
p-phenylenediamine (TMPD). He also obtained good results for the
ICT of cyto c oxidase using TMPD. The 2,6-dichlorophenolindo-
phenol has been examined thoroughly in the OTTLE cell and used
for potentiometry of cyto c by Heineman (21). Although only a few
naphthaquinones appear in the table, several other naphthaquinones, particularly those substituted appropriately for greater
water solubility, are being characterized. These naphthaquinones
will serve as M-T's in the potential range of +100 to +200 mV.
Phenazine metho- and etho-sulfate are well known mediators which
have been widely used in potentiometric titrations of biocomponents. The pteridones have been suggested for coupling to NAD^+
reduction by Kwee and Lund (33). We hope that further work will
be forthcoming from their laboratory utilizing these pteridones.

Acknowledgement

The financial support provided by NSF Grant MPS 73-04882 and
NIH-PHS Grant No. GM 19181 is gratefully acknowledged.

Literature Cited

1. Hawkridge, Fred and Kuwana, Theodore, Anal. Chem., (1973)
    45, 1021.
2. Heineman, William and Kuwana, Theodore, Acc. Chem. Res.,
    (1976) 9, 241.
3. Heineman, William and Kuwana, Theodore, Biochem. Biophys.
    Res. Commun., (1973) 50, 892.
4. Mackey, L.N., Kuwana, T., and Hartzell, C.R., FEBS Lett.,
    (1973) 36, 326.
5. Rodkey, F.L. and Donovan, J.A. Jr., J. Biol. Chem., (1959)
    234, 677.

6. Thevenot D. and Leduc, P., 3rd International Symposium on Bioelectrochemistry, Juelich (1975).
7. Ke, B. and Hawkridge, F.M., unpublished results.
8. Steckan, Eberhard and Kuwana, Theodore, Ber. Bunsenges. Phys. Chem., (1974) 78, 253.
9. Mackey, L.N. and Kuwana, Theodore, 3rd International Symposium on Bioelectrochemistry, Juelich (1975).
10. Szentrimay, Robert, (1976) Ph.D. Thesis, Ohio State University.
11. Hennig, Horst and Gürtler, Oswald, J. Organometal. Chem., (1968) 11, 307.
12. Mason, J.G. and Rosenblum, Myron, J. Am. Chem. Soc. (1960) 82, 4206.
13. Gubin, S.P. and Perevalova, E.G., Dokl. Akad. Nauk SSSR, (1962) 143, 1351.
14. Perevalova, E.G., Gubin, S.P., Smirnova, S.A. and Nesmeyanov, A.N., Dokl. Akad. Nauk. SSSR (1964) 155, 857.
15. Yu, C.A. and Yu, Linda, Biochem. Biophys. Res. Commun., (1976) 70, 1115.
16. Fujihira, Y., Kuwana, T. and Hartzell, C.R., Biochem. Biophys. Res. Commun., (1974) 61, 488.
17. Yeh, P. and Kuwana, T., J. Electrochem. Soc., (1976) 123, 1334.
18. Penden, A.A., Leont'evskaya, P.K., L'vova, T.I. and Nikolskii, B.P., Dokl. Akad. Nauk. SSSR, (1969) 189, 115.
19. Strojek, J.W. and Kuwana, T., Electroanalytical Chemistry and Interfacial Electrochemistry, (1968) 16, 471.
20. Penden, A.A., Zakharevskii, M.S. and Leont'evskaya, P.K., Kinetika: Kataliz, (1966) 7, 1074.
21. Heineman, W.R., Norris, B.J. and Goelz, J.F., Anal. Chem., (1975) 47, 79.
22. Caughey, W.S., Wallace, W.J., Volpe, J.A. and Yoshikawa, S., in "The Enzymes" (P.D. Boyer ed.) Volume XIII Part C, p. 299, Academic Press, New York, 1976.
23. Schroedel, Nancy, (1976) Ph.D. Thesis, The Pennsylvania State University.
24. Reinhammar, Bengt R. M., Biochimica et Biophysica Acta, (1972) 275, 245.
25. Pecht, Israel, Israel Journal of Chemistry, (1974) 12, 351.
26. Mendez, Hernandez and Lucenta, F., An. Quim, (1968) 64, 71.
27. Mendz, Hernandez, J. Acta Salmanticensia, Cienc (1967-1968) 33, 41.

28. Cordova-Orellana, Lucena-Conde, F., Talanta (1971) 18 505.
29. Laitinen, H.A. and Conley, J.M., Anal. Chem., (1976) 48, 1224.
30. Kolthoff, I.M. and Tomsicek, Wm. J., J. Phys. Chem., (1936 40, 247.
31. Malik, W. and Ali, S.I., Indian J. Chem., (1963) 1, 374.
32. Nickolls, P. and Chance, B., in "Molecular Mechanisms of Oxygen Activation", (O. Hayaishi, ed.) p. 479 Academic Press, New York 1974.
33. Kwee, S. and Lund, H., Bioelectrochemistry and Bioenergetics, (1975) 1, 137.
34. Schilt, A.A., Anal. Chem., (1963) 35, 1599.
35. Miller, R.R., Brandt, W.W. and Puke, M., J. Am. Chem. Soc., (1955) 77, 3178.
36. Dwyer, F.P. and McKenzie, H.A., J. Proc. Roy. Soc. N.S. Wales, (1947) 81, 93.
37. Fortune, W.B. and Mellon, M.G., Ind. Eng. Chem., Anal. Ed., (1938) 10, 60.
38. Harvey, A.E. and Manning, D.L., J. Am. Chem. Soc., (1952) 74, 4744.
39. Moss, M.L. and Mellon, M.G., Ind. Eng. Chem., Anal. Ed., (1942) 14, 862.
40. Schilt, A.A., "Analytical Applications of 1,10-Phenanthroline and Related Compounds", Pergamon Press, New York, (1969).
41. Dywer, F.P., Gibson, N.A. and Gyarfas, E.C., J. Proc. Roy. Soc. N.S. Wales, (1942) 84, 80.
42. Burstall, F.H., Dwyer, F.P. and Gyarfas, E.C., J. Chem. Soc., (1950), 953.
43. (a) Volke, J., Collect. Czechoslov. Chem. Commun., (1968) 33, 3044.
    (b) Volke, J. and Volkova, V., Collect. Czechoslov. Chem. Commun., (1969) 34, 2037.
44. Kolthoff, I.M. and Tomsicek, W.J., J. Phy. Chem., (1935) 39, 945.
45. Michaelis, L. and Hill, E.S., J. Am. Chem. Soc., (1933) 55, 1481.
46. Albrecht, A.C. and Simpson, W.T., J. Am. Chem. Soc., (1955) 77, 4455.
47. Dutton, P.L., Wilson, D.F. and Lee, C.P., Biochem., (1970) 9, 5077.
48. Dickens, F. and McIlwain, H., Biochem. J., (1938) 32, 1615.

49. Friedheim, E.A.H., Biochem. J., (1934) 28, 180.
50. Friedheim, E.A.H. and Michaelis, L., J. Biol. Chem. (1931) 91, 355.
51. Kwee, S. and Lund, H., Biochim. Biophys. Acta, (1973) 297, 285.
52. Fieser, L.F. and Fieser, M., J. Am. Chem. Soc., (1934) 56, 1565.
53. Conaut, J.B., Kahn, H.M., Fieser, L.F. and Kurtz, S.S., J. Am. Chem. Soc., (1922) 44, 1382.
54. Michaelis, L., Schubert, M.P. and Granick, S., J. Am. Chem. Soc., (1939) 61, 1981.

10

Rotating Ring Disk Enzyme Electrode for Biocatalysis Studies

RALPH A. KAMIN, FRANK R. SHU,[1] and GEORGE S. WILSON
Department of Chemistry, University of Arizona, Tucson, Ariz. 85721

In recent years there has been considerable interest in catalytic surface reactions particularly those of biological interest. This has been manifested, for example, in the rapid development of immobilized enzyme technology (1,2) and electrochemical sensors based on electroactive product formation within an enzyme layer (3,4). In comparing the kinetic behavior of an immobilized enzyme with its soluble counterpart, it is necessary to establish that the overall reaction rate is catalysis rather than mass transport limited. It has been shown, for example, that immobilized enzymes in flowing streams give apparent Michaelis constants K_M', that are flow rate dependent (5). Under conditions where the overall reaction is limited by mass-transport supply of substrate to the catalytic surface, K_M' is larger than expected. One is then tempted to conclude that the properties of the enzyme have been modified by immobilization. On the contrary, increasing flow (mass transport) rates may lead to a limiting value for K_M' essentially identical to that of the soluble enzyme (6).

The rotating disk electrode as described by Levich (7) appears to offer an experimentally facile means for varying the rate of substrate mass transport. The addition of a concentric ring (rotating ring disk electrode) (8) permits independent monitoring of the reaction at the disk surface. We have recently (9) derived the theory describing the response of the rotating disk enzyme electrode. In the present work we report further experimental studies in support of this theoretical model. The system selected for study is the glucose/glucose oxidase reaction:

$$\text{Glucose} + O_2 \xrightarrow[\text{oxidase}]{\text{glucose}} \text{Gluconic Acid} + H_2O_2 \qquad (1)$$

The peroxide produced is either monitored directly or coupled

[1] Present address: Smith-Kline Instruments, 880 W. Maude Ave., Sunnyvale, CA 94086

with the indicator reaction:

$$H_2O_2 + 2H^+ + 2I^- \xrightarrow{\text{molybdate}} I_2 + 2H_2O \qquad (2)$$

Experimental

Instrumentation. The four electrode potentiostat used in these studies was similar to that described by Shabrang and Bruckenstein (10). The rotating disk electrode, Model DT-6, was purchased from the Pine Instrument Co., Grove City, PA. The disk was a 0.5 cm deep cavity with a radius of 0.382 cm according to the manufacturer's specifications. When filled with carbon paste the calculated disk area was 0.46 cm^2. The width of the platinum ring electrode was 0.024 cm and was separated from the disk by a 0.016 cm wide epoxy gap. The collection efficiency measured experimentally in the usual way as a carbon paste electrode (8) was 0.18 and was in good agreement with experimental results. A platinum wire counter electrode and a Ag/AgCl reference electrode (E°' = 0.200 v) were employed. A Pine Instruments Model PIR rotator was used to control electrode rotation speed.

Preparation of Glucose Oxidase Electrode. The carbon paste was prepared in the usual manner from 5 g of graphite powder #38 (Fisher Scientific Co.) and 3 ml of Nujol except that 10 mg (except where otherwise specified) of n-octadecylamine (technical grade, Aldrich) was also added. The carbon paste was packed firmly into the disk cavity of the DT-6 electrode which was then polished with a piece of weighing paper. After the ring and gap were carefully cleaned, the electrode was allowed to rotate in a 12.5% glutaraldehyde solution for 10 - 15 min. followed by a 1 minute washing with cold 0.2M phosphate buffer pH 6.5. (Glutaraldehyde must be freshly purified and stored below 0°C as it readily polymerizes (11)). The rotating electrode was dipped into a bovine serum albumin solution (0.1 g/ml) (BSA Fraction V 96-99%, Sigma Co.). After 2 - 3 minutes the electrode was washed for 1 minute in cold phosphate buffer. The electrode was then removed from the rotator and positioned with the electrode surface facing up. A glucose oxidase solution prepared by dissolving 0.3 g of the enzyme (Glucose Oxidase E.C. 1.1.3.4 Sigma Type II 15,000 units/g) in 1 ml. of 5% glutaraldehyde solution (buffered with phosphate at pH 6.5) was applied to the disk surface. After standing at room temperature for 5 min., the excess enzyme solution was discarded and the gap and ring were carefully cleaned. Rotating the electrode in cold phosphate buffer at 2500 rpm for 5 min. aids in removing physically entrapped or weakly bonded enzyme. When not in use the electrode was stored in phosphate buffer at 5°C.

Solutions and Reagents

Unless otherwise mentioned, all chemicals used were reagent grade. The stock solution of 0.1 M glucose was allowed to muta-rotate at room temperature for at least 24 hr. before using. When the course of the reaction was measured by following I_2 formation (Reaction 2) a KI-buffer catalyst described previously (12) was employed. Where direct monitoring of peroxide forma-tion (Reaction 1) is possible the glucose is dissolved in a 0.05 M phosphate buffer pH 6.5.

Procedure. The enzyme electrode was allowed to rotate for about 30 sec. in the glucose solution at which time a potential was applied to the respective indicating electrode. The iodine formed in Reaction 2 was monitored at the disk by applying a potential of -0.2 V vs Ag/AgCl reference. Direct peroxide for-mation (no iodide present) was monitored at the platinum ring by holding the potential at -0.2 V followed by a step to 0.75 V at which point the current transient was measured. The potential was then returned to -0.2 V until the next measurement.

Enzyme Electrode

Theoretical Model. The details of the digital simulation calculations for this electrode have been presented elsewhere (9). Our model assumes the existence of an enzyme layer ex-tending into solution from the electrode surface (X=0). This uniformly distributed thin enzyme layer is assumed not to inter-fere with diffusion of species to or from the electrode surface. The enzyme layer lies within the minimum hydrodynamic layer justifying the assumption that solution flow in the electrode vicinity is also unaffected by the immobilization process. Michaelis-Menten kinetic theory is assumed to describe the en-zymatic reaction.

Figure 1 illustrates the nature of the concentration grad-ients at the electrode surface for a particular set of condi-tions. The steady state product (or coupled product) concentra-tion gradient is first simulated for the rotating electrode at open circuit. Product concentration increases as substrate pen-etrates the enzyme layer from the solution side. If a potential is applied to the disk in a region where the product is electro-active, its concentration at the electrode surface drops to zero. Eventually the steady-state condition shown in Figure 1 is at-tained. It will be noted that the concentrations in the outer portion of the enzyme layer are relatively unaffected by the po-tential perturbation.

The rate of product formation is given by Michaelis-Menten theory

$$\frac{d\,[P]}{dt} = k_3 C_E / (K_M / [S] + 1) \qquad (3)$$

where k_3 is the rate constant for the irreversible conversion of the enzyme-substrate complex into products and K_M the Michaelis constant. C_E in this case is the analytical concentration of active enzyme in the immobilized layer and S the substrate concentration in the enzyme layer.

In order to evaluate the relative effects of catalysis and convective mass transport a reaction velocity parameter, V, is defined:

$$V = k_3 C_E t_k / K_M \qquad (4)$$

The convection time constant, t_k, has been derived previously by Prater and Bard (13) and is given by

$$t_k = (0.51)^{-2/3} D^{-1/3} \nu^{1/3} \omega^{-1} \qquad (5)$$

where ν is the kinematic viscosity (cm^2/sec) and ω the rotation speed in rad/sec. For a given enzyme electrode, V reflects the amount of product formed in a given time and is dependent only on ω, to which it is inversely proportional. For large values of V e.g. $V \geq 10$ the catalysis rate is extremely fast and the overall reaction becomes convection mass transport limited. For $V \leq 0.1$ the enzymatic reaction is catalysis rate limited. Thus, by varying the electrode rotation speed, the flux of substrate can be modulated to change the nature of the rate limiting process. The ratio C/K_M where C is the bulk substrate concentration also serves to define the current response. We have also shown (9) that an optimal rotation speed for current measurement will result from increased substrate mass transport on one hand and decreased product production due to short contact time with the catalytic layer on the other. The steady state current relationships are presented below:

Case I - Mass Transport Limited Rate ($V \geq 10$)

From simulation it can be shown (by analogy to a Lineweaver-Burk plot (14)):

$$\frac{nFAdk_3 C_E}{i_s} = 1.22 \left[\frac{d^2}{Dt_k}\right]^{1/2} \left[\frac{k_3 C_E t_k}{C}\right] + b \qquad (6)$$

where i_s is the steady state current at the disk, d is the enzyme layer thickness; b is a function only of ω and D. All other parameters have the usual electrochemical significance. At low substrate concentrations the first term of Equation 6 is much greater than b and the steady-state current becomes

$$i_s = 0.65 nFAD^{2/3} \nu^{-1/6} \omega^{1/2} C \qquad (7)$$

which is practically identical to the Levich equation (7) for a rotating disk electrode as expected.

Case II - Catalysis Limited Rate (V≤0.1)

An expression analogous to Equation 6 can be derived for this case:

$$\frac{nFAdk_3 C_E}{bi_s} = \frac{K_M}{C} + 1 \tag{8}$$

For high substrate concentrations (C>>K_M) Equation 8 can be written

$$i_{max} = \frac{1}{b} nFAdk_3 C_E \tag{9}$$

It can be seen that an experimental plot of i_{max}/i_s vs C^{-1} for the relationship

$$\frac{i_{max}}{i_s} = \frac{K_M}{C} + 1 \tag{10}$$

will yield the Michaelis constant.

Characteristics of the Immobilized Enzyme Layer

Figure 2 gives a rough schematic representation of the immobilized layer. The presence of the amine dissolved in the Nujol and/or adsorbed on the graphite particles appears to be essential to the formation of a stable enzyme layer of high biological activity. If product formation is to be monitored at the disk then an optimal amount of amine must be used. If the concentration is too low, then insufficient coupling sites are available whereas too much amine causes current suppression. At the amine level suggested above the electrochemistry of I^- (cyclic voltammetry and RDE limiting currents) are virtually identical in the presence or absence of the amine in the carbon paste. Based on the Levich Equation (7) the effective electrode area is approximately 50% of the projected area suggesting that about half of the graphite is in contact with the solution at the electrode surface. The BSA added in the next step also appears to be essential to enzyme layer stability. It probably functions along with the glutaraldehyde as a multi-functional cross-linking reagent. Limiting disk currents for I_3^- reduction are identical for a pure carbon paste electrode and an enzyme electrode (no substrate present) suggesting that the immobilization process does not interfere with diffusion controlled reduction of the electroactive species.

Little is known at present about the micro environment of the immobilized enzyme active site. The pH optimum is shifted

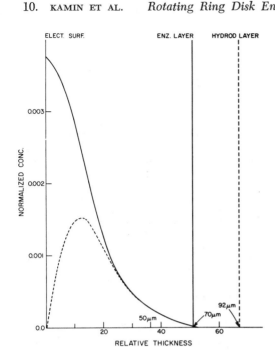

Figure 1. *Simulation of steady state product concentration gradients conditions* $CKM = 3.334$, $V = 0.0113$, $d = 0.007$ cm, and $\omega = 1600$ rpm. *(———), disk profile at open circuit; (———), profile for disk electrode polarized at product potential.*

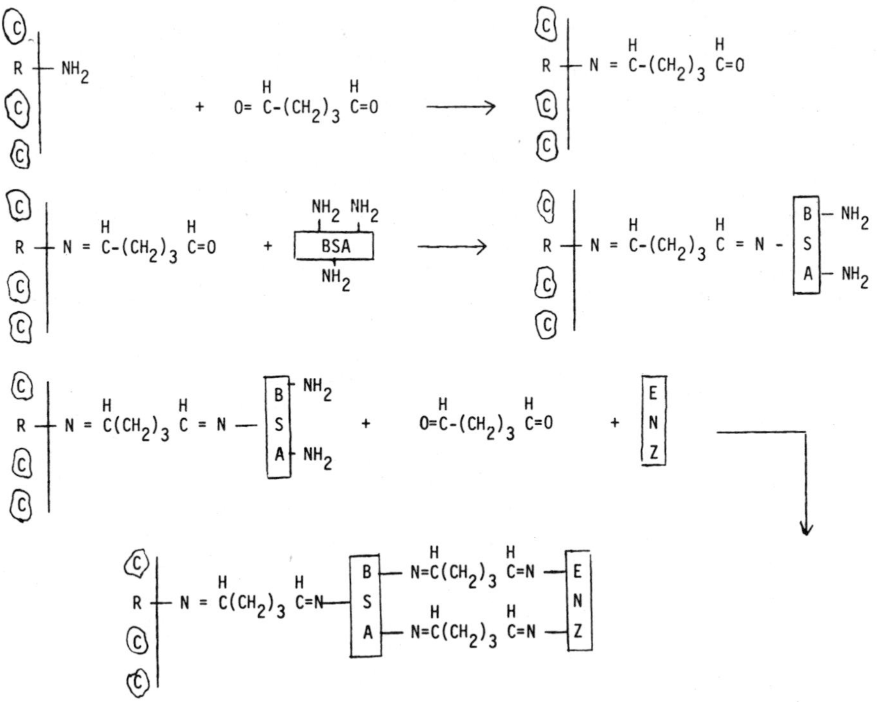

Figure 2. *Scheme for enzyme layer formation*

to about 6.5 which is around one pH unit higher than the soluble enzyme (6). Such alkaline shifts may be caused by a reduction of positive charge on the enzyme due to the coupling process. This would occur as positively charged ε-amino lysyl or guanido groups react with the glutaraldehyde (15). The Michaelis constant might be affected in an analogous manner by the electrostatic potential of the active site microenvironment for cases of charged substrates. The possibility of varying this potential by changing the potential applied to the disk is an attractive one, especially in view of present electrochemical methodology.

When properly immobilized, a stable catalytic layer is obtained. After an initial decay in activity, probably due to non-bonded enzyme, the activity decays slowly. We have observed only 15% decay after 30 days and have used electrodes for glucose determinations after two months of regular use. It is not known at present whether loss of activity is due to enzyme layer "washout", enzyme denaturation, or possibly both.

Use of the Ring for Reaction Monitoring

There are some important advantages in using the ring for monitoring the course of the enzymatic reaction. First, the electrode to which the enzyme is attached may not be suitable or optimal for monitoring product formation. We have not been able to obtain reproducible results for peroxide when monitored directly at the rotating disk whereas the platinum ring is quite suitable. Second, the course of the reaction can be measured without potential perturbation of the catalytic surface or alternatively at a potential at which neither products nor reactants are electroactive. Finally, chemical and electrochemical characteristics of reaction products can be determined in a manner analogous to a conventional rotating ring disk electrode.

Figure 3 demonstrates the manner in which the ring "tracks" the disk as the rate of the enzymatic reaction is varied by changing substrate concentration. The slope of Figure 3 (equal to 0.4) is independent of ω over a wide range. In this case the ring current is measured with the disk at open circuit. If a potential characteristic of I_3^- reduction is applied simultaneously to both ring and disk an "apparent collection efficiency" of 0.30 is obtained. This is only a 25% reduction over the conditions of Figure 3 suggesting that a large portion of the product which is produced in the enzyme layer can be collected after it diffuses back into the solution.

The direct proportionality between the disk and ring currents over a wide range of conditions makes possible the evaluation of enzyme kinetic parameters at the ring. Using Equation 10 and the steady-state ring current, a K_M of 9.59 mM was obtained. This is in good agreement with values obtained at the disk using Reaction 2 (9) as well as with previously reported values for

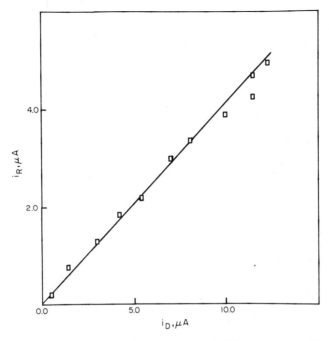

Figure 3. Relationship between ring and disk current $\omega = 400$ rpm. Ring current measured with disk at open circuit.

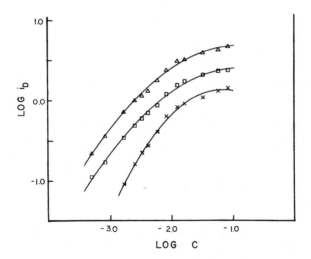

Figure 4. Experimental steady-state ring current $[I_R(\mu A)]$ as a function of glucose concentration (M) at various rotation speeds; 400 rpm ($\triangle$), 900 rpm ($\square$), and 1600 rpm ($\times$)

both the soluble and immobilized enzyme (<u>1</u>).

In Figure 4 the relationship between substrate concentration and steady-state ring current is demonstrated. As in the case of the disk, the current is observed to decrease as the rotation speed is increased. At 1600 rpm, however, the greatly enhanced substrate mass transport results in higher <u>sensitivity</u> at the expense of decreased linear range and absolute current.

When a potential is applied to the ring a current transient results which decays to steady-state in about 60 sec. It is possible to shorten the measurement time and improve sensitivity by measuring the current before steady-state is reached. As expected, the current at time t is linearly proportional to glucose concentration and a sensitivity improvement of at least 10% over steady-state is observed without significant deterioration of precision (better than 2%).

Conclusion

The possibility of producing a stable catalytic surface attached to an electrode has been demonstrated. Using a rotating ring disk enzyme electrode, the effects of substrate mass transport and kinetic control of surface catalyzed reactions can be studied. The use of the ring makes possible the independent monitoring of the reactions occurring at the disk. The evaluation of kinetic parameters has been demonstrated paving the way for detailed electrochemical characterization of immobilized biosurfaces.

Acknowledgement

This work was supported in part by National Science Foundation Grant CHE 73-08683-A03.

Literature Cited

1. Weetall, H. H., Anal. Chem., (1974), <u>46</u>, 602-615A.
2. Wiseman, A., ed., "Handbook of Enzyme Biotechnology," Halsted Press, New York (1975).
3. Mell, L. D. and J. T. Maloy, Anal. Chem., (1975), <u>47</u>, 299-307.
4. Guilbault, G. G. and G. J. Lubrano, Anal Chim. Acta, (1973) <u>64</u>, 439-455.
5. Hornby, W. E., M. D. Lilly, E. Crook and P. Dunnill, Biochem J., (1968) <u>107</u>, 669-674.
6. Smith, G. L., M. S. Thesis, University of Arizona (1975).
7. Levich, V. J., "Physicochemical Hydrodynamics," Prentice-Hall, Englewood Cliffs, N. J. (1962).
8. Albery, W. J. and M. L. Hitchman, "Ring-Disk Electrodes", Oxford University Press, London (1971).

9. Shu, F. R. and G. S. Wilson, Anal Chem., (1976), <u>48</u>, 1679-1686.

10. Shabrang, M. and S. Bruckenstein, J. Electrochem Soc., (1975), <u>122</u>, 1305-1311.

11. Hadju, J. and P. Friedrich, Anal. Biochem, (1975), <u>65</u>, 273-280.

12. Malmstadt, H. V. and H. L. Pardue, Anal Chem, (1961), <u>33</u>, 1040-1047.

13. Prater, K. B., and A. J. Bard, J. Electrochem Soc., (1970), <u>117</u>, 207-213.

14. Mahler, H. R. and E. H. Cordes, "Biological Chemistry," Harper and Row, New York (1966).

15. Goldstein, L. in "Methods in Enzymology" Vol. 19, G. E. Perlmann and L. Lorand, Eds. Academic Press, New York, (1970), pp. 935-978.

11

Electrokinetic Potentials in a Left Ventricle/Aorta Simulator

EUGENE FINDL and ROBERT J. KURTZ

ARK Research, Farmingdale, N.Y. 11735

There are two terms in electrochemistry that sound alike but cover different aspects of the field, electrode kinetics and electrokinetics. The first term is used to describe electrode reaction rates. The second term (our subject herein) is used to describe interfacial electrochemical phenomena observed when an electrolyte and a solid surface move with respect to each other.

Electrokinetic phenomena are classically divided into four categories, i.e., electrophoresis, electro-osmosis, sedimentation potentials and streaming potentials. There are however, several lesser known electrokinetic effects, namely moto-electric effects, Ueda effects and acousto-electric effects. Of these electrokinetic phenomena, streaming potentials, under turbulent flow conditions, are theorized to contribute significantly to the electrical signals attributed to mammalian hearts.

It is well known that blood flows through the heart and certain blood vessels under turbulent flow conditions(1). Therefore, based upon the premise that streaming potentials of significant magnitude can be generated by such flow conditions, we decided to investigate the possibility that streaming potentials contribute a substantial part of the potentials seen in an electro-cardiogram (EKG).

The work of Miller and Dent(2) is of particular interest in this regard. They presented experimental evidence, both in vitro and in vivo (with dogs), that streaming potentials are the cause of at least the T wave portion of the EKG. Others(2-6) have reported the existence of streaming potentials in cardio-vascular components.

180

Our approach to the problem of demonstrating
that the EKG is at least partially due to streaming
potentials has been to first demonstrate in vitro,
with mechanical left ventricle simulators, that an
EKG like signal can be generated. Second, to
demonstrate that EKG like signals can be generated
in vivo using a mechanical, pulsatile, heart pump.
Some of the experimental results of our first step
are presented herein.

Experimental Apparatus

Our latest left ventricle simulator (LVS III)
is shown on Figure 1. [Two other models, LVS I
and LVS II, were of similar design.] A dc motor
driven cam moves a shaft that displaces a rubber
diaphragm. As the diagram is displaced inward,
fluid in the "left ventricle" chamber is forced out,
through a one way rubber valve and into a
distensible balloon (aorta). During the ventricular
cycle, when compression occurs, the "aorta" expands
due to the outward flow of fluid. When the
diaphragm is brought back to its starting position,
fluid from the "aorta" flows back into the "left
ventricle" chamber through a second one way valve.
Stroke volumes were varied between 20 and 50 ml by
varying the flow restriction caused by the check
valves.

Electrodes (Ag/AgCl) were placed at various
locations on the "left ventricle." The electrodes
were placed out of the flowing streams to minimize
moto-electric effect artifacts. [More will be
said about this problem area later.] Coaxial
cable leads were adapted to compression fittings
to connect the electrodes to the readout
instrumentation.

Simulator electrodes were numbered as shown on
Figure 2. The LVS II version had an additional
electrode (#6) placed in the body of the LVS.
Water bath electrodes (Figure 3) were labeled A,
B, C and D. These were located as follows: A, near
one valve; B, 2-6 cm from A; C, 2-6 cm from D; D,
near the other valve.

The simulators were made ionically conductive
by drilling many small holes into the left
ventricle chamber and the valve holder sections.
Bulk transfer of fluid from the LVS's and the
water bath into which they were submerged was
prevented by a coating over the holes. Several
different coatings were evaluated. Among the

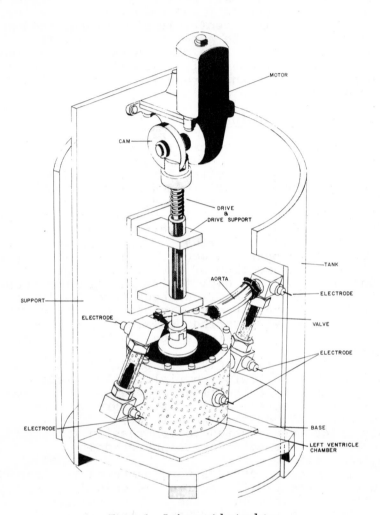

Figure 1.   Left ventricle simulator

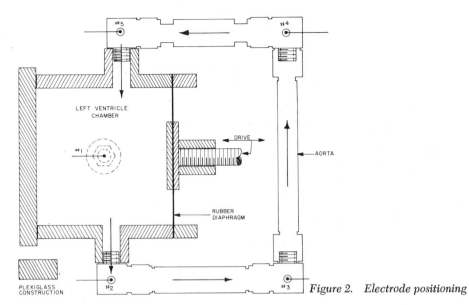

Figure 2.   *Electrode positioning*

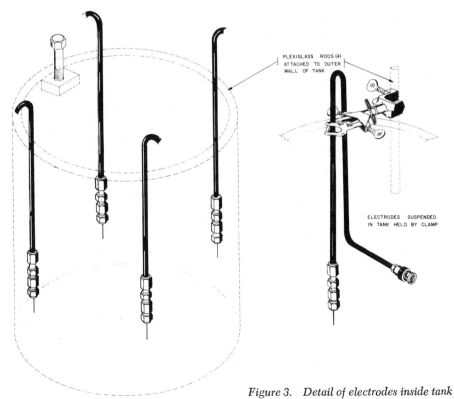

Figure 3.   *Detail of electrodes inside tank*

materials were a thin latex, a cellulose acetate
reverse osmosis membrane, collagen, collodion and
a cellulose battery separator material. Of these,
collodion and the cellulose membranes appeared to
be the most satisfactory. The others were either
too resistive or too difficult to mechanically
attach to the LVS's.
       The left ventricle was immersed in a water
bath to simulate the "aqueous environment" of a
mammalian body. Figure 3 illustrates how four
Ag/AgCl electrodes were attached to the bath.
       A Tektronix 7514 dual channel, storage
oscilloscope with 7A22 differential amplifiers was
used to monitor the electrokinetic potentials.
Input impedance was increased to 10^{14} ohms by
means of solid state voltage followers. In some
tests, in addition to the potentials generated,
we also monitored the "left ventricular" pressure,
using a strain gauge transducer.

Test Results

       Examples of the electrokinetic potentials
generated by the simulators is shown on Figures 4,
5, 6 and 7. All tests were run at ambient
temperature, using various concentrations of NaCl
as the electrolyte.
       Figure 4 illustrates the waveforms obtained
from various combinations of electrodes using the
LVS II. The upper traces are the electrokinetic
potentials while the lower trace illustrates the
left ventricular pressure. Electrokinetic
potentials were measured having the following
vertical deflection scale factors,

Electrode Pair	Scale (mv/div)	Electrode Pair	Scale (mv/div)
$1^-, 2^+$	50	$3^-, 4^+$	20
$1^-, 3^+$	200	$3^-, 5^+$	200
$1^-, 4^+$	200	$3^-, 6^+$	200
$1^-, 5^+$	20	$4^-, 5^+$	200
$1^-, 6^+$	20	$4^-, 6^+$	200
$2^-, 3^+$	200	$5^-, 6^+$	20
$2^-, 4^+$	200		
$2^-, 5^+$	20		
$2^-, 6^+$	50		

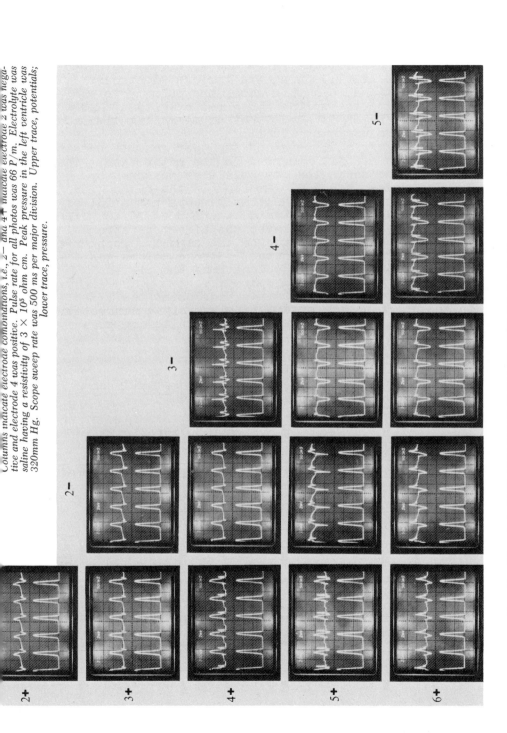

Columns indicate electrode combinations, i.e., 2− and 4+ indicate electrode 2 was negative and electrode 4 was positive. Pulse rate for all photos was 66 P/m. Electrolyte was saline having a resistivity of 3 × 10⁵ ohm cm. Peak pressure in the left ventricle was 320mm Hg. Scope sweep rate was 500 ms per major division. Upper trace, potentials; lower trace, pressure.

The LVS was not immersed in the water bath. Pulse
rate was 66 pulses per minute (P/M). The
electrolyte was saline having a resistivity of
3×10^5 ohm cm. Peak ventricular pressure was
320 mm Hg.

Figure 5 illustrates the waveforms obtained
from the LVS III using its various electrode
combinations. In this case, the pulse rate was
76 P/M and the saline had a resistivity of 1,040
ohm cm, which is about that of mammalian tissue.
Stroke volume was ≈ 50 ml.

Figure 6 illustrates the test results obtained
with the LVS III immersed in the water bath. The
top trace shows the potentials as measured by
electrodes (A,B,C,D) immersed in the bath. The
bottom trace shows the simultaneous potential of
electrodes 1 and 2, inside the LVS.

Figure 7 illustrates the effect of stroke
volume on the amplitude of the signal. Pulse rate
was 72 P/M for all photos. Stroke volume was
varied by changing the length of the piston
pushing the LVS III diaphragm.

A series of tests was made using the LVS III
to determine the effect of electrolyte conductivity
on signal voltage level. The results are shown
on Table 1.

Table 1

Effect of Electrolyte Conductivity on
Signal Voltage Level

Electrolyte Conductivity $(ohm^{-1}.cm^{-1})$	Signal Voltage Level Peak (mV)	Signal Voltage Level Average (mV)
$1.1 \cdot 10^{-2}$	0.5	0.12
$2.0 \cdot 10^{-3}$	3	0.60
$9.1 \cdot 10^{-4}$	5	0.87
$2.9 \cdot 10^{-4}$	12	2.40
$1.5 \cdot 10^{-4}$	18	2.95
$1.0 \cdot 10^{-4}$	20	3.82
$3.0 \cdot 10^{-5}$	40	7.88
$2.2 \cdot 10^{-5}$	44	9.24
$1.4 \cdot 10^{-5}$	47	10.86

Theoretically, streaming potential is inversely
proportional to conductivity. Figure 8 illustrates
that our experimental results approximate the theoretical values over the range 10^{-2} to 10^{-5} $ohm^{-1}cm^{-1}$.

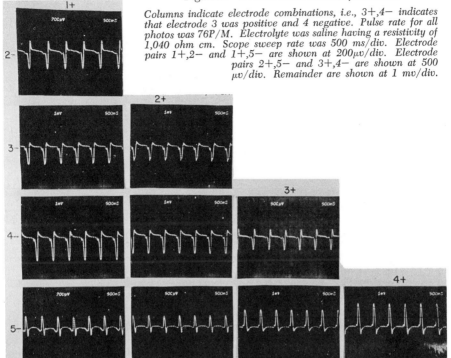

Figure 5. LVS III electrokinetic potentials.

Columns indicate electrode combinations, i.e., 3+,4− indicates that electrode 3 was positive and 4 negative. Pulse rate for all photos was 76P/M. Electrolyte was saline having a resistivity of 1,040 ohm cm. Scope sweep rate was 500 ms/div. Electrode pairs 1+,2− and 1+,5− are shown at 200μv/div. Electrode pairs 2+,5− and 3+,4− are shown at 500 μv/div. Remainder are shown at 1 mv/div.

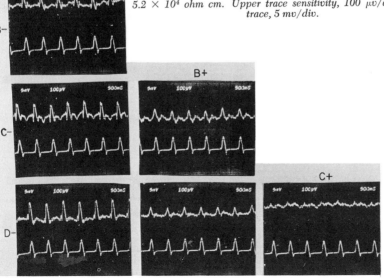

Figure 6. LVS III wave forms.

Columns indicate upper trace electrode combinations. Lower trace was 1+,2− electrode pair. Pulse rate, 80 P/M. Resistivity— 5.2×10^4 ohm cm. Upper trace sensitivity, 100 μv/div.; lower trace, 5 mv/div.

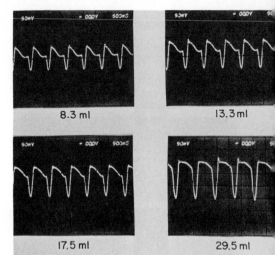

Figure 7. Effect of stroke volume on signal amplitude. Electrodes 4+,5− of the LVS III. Average stroke volumes are given below photos.

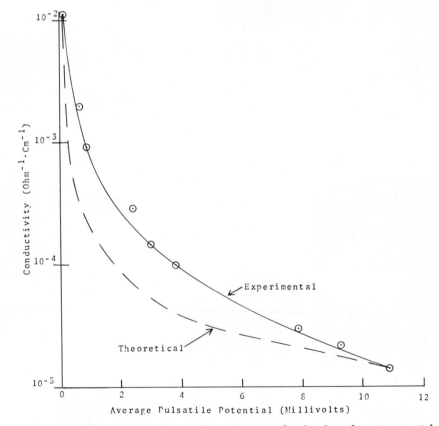

Figure 8. Effect of electrolyte conductivity on pulsatile electrokinetic potential (electrode 4 and 5; LVS III)

Discussion

Historically, streaming potentials were first described by Quincke in 1859.(7) Other 19th century investigators included Zollner(8), Edlund(9), Haga(10) and Clark(11). Helmholtz(12) added a theoretical basis for streaming potential, incorporating his famous double layer theories into the explanation. Smoluchowski refined Helmholtz's theories in her treatise published in 1921. Most recent effort in the field has been concerned with the determination of zeta potentials by using streaming potential measurements.

With but few exceptions(11,13,14) the experimental effort described in the streaming potential literature has dealt with laminar flow through capillary tubes or porous plugs. The region of turbulent flow has been largely ignored until recently. Kurtz, Findl, Kurtz and Stormo(15) extended the region of streaming potential measurements well into the turbulent region, utilizing tubing up to 3.8 cm in diameter. Further, they extended the basic streaming potential relationships well into the turbulent flow region.

Laminar region

$$E = 8[D\zeta L/\pi d^2 k]u$$

Turbulent region

$$E = [0.04(\rho/\mu)^{0.75}][D\zeta L/\pi d^{1.25}k]u^{1.75}$$

where d = tubing diameter   u = fluid velocity
      D = dielectric constant ζ = zeta potential
      E = streaming potential μ = fluid viscosity
      k = fluid conductivity ρ = fluid density
      L = electrode spacing

The major difference between the two relationships is that in the turbulent region, as compared to the laminar region, the streaming potential increases as a function of velocity to the 1.75 power rather than linearly. Further, the potential diminishing effect of enlarging tubing diameter is much less in the turbulent region, i.e., d^{-2} in laminar versus $d^{-1.25}$ in turbulent. Overall, the streaming potential increases much more rapidly in the turbulent region than would have been expected using extrapolations from the laminar flow region.

There are two other electrokinetic effects
that might be capable of generating potentials such
as were illustrated. These are moto-electric
effects(16,17,18) and Ueda effects(19,20). Unlike
streaming potentials, these two effects are due to
the interaction between an electrode and an
electrolyte moving relative to it. [Streaming
potentials on the other hand, are due to the
relative motion between an electrolyte and a solid
surface, generally an insulator. The potential is
measured by a pair of electrodes which are not
located in the moving electrolyte.]

Moto-electric effects are readily distinguish-
able by the slow response of this potential to
changes in fluid velocity. Further, reversal of
fluid direction does not cause immediate reversal
of polarity of the potential as is the case with
streaming potentials. In general, moto-electric
effects simply cause a shift in baseline and do
not contribute to pulsatile potentials.

Ueda effects are essentially due to rapid
vibratory movement of an electrode-electrolyte
interface. This vibrating motion results in a
sinusoidal, alternating potential being developed.
It is typically caused by a mechanical vibration of
the electrolyte, such as caused by banging a water
bath with a hard object, or by rapidly closing a
valve causing a water hammer in a pipe. [Ultra-
sonic acousto-electric effects such as those
described by Yeager et al.(20,21,22) are of too
high a frequency to be factors in our experimental
results.]

We have investigated the possibility that Ueda
effects were contributing to the potentials
generated by the LVS's. When electrolytes of
high resistivity (> 10^4 Ω cm) were used, hitting
the LVS with a metal object did produce measurable
sinusoidal voltages. With electrolytes of lower
resistivity, the effect was less significant. The
Ueda effect does not account for the pulsatile
potentials measured in the water bath, but it did
contribute a "sinusoidal noise" signal due to wave
motions in the bath.

A factor that indicates that it was indeed
streaming potentials that we measured was that the
wave shape was dependent upon the rubber check
valves. These valves were individually cast in
our laboratory by hand, using a surgical latex.
Each had a character of its own. As a result, the
opening and closing characteristics of each

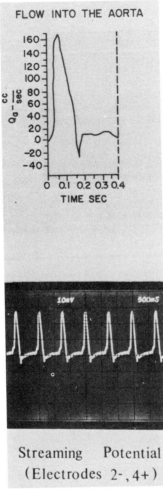

Figure 9. *Wave form comparison*

differed. This resulted in a noticeable variation
in signal waveform. It is much more reasonable to
assume these variations in signal waveforms are
due to flow variations and thus streaming potentials
rather than due to acoustic effects causing Ueda
potentials.

As a final indication that the potentials
measured were indeed streaming potentials, Figure 9
illustrates a comparison between the wave shape of
electrodes 4 and 5 and the flow of blood into the
human aorta(1). Note the close similarity of wave
form.

The potentials measured using the LVS's with
normal saline, were an order of magnitude lower
than those typically obtained in vivo, using
mammals. It was not our objective to duplicate
in vivo surface-electrolyte conditions or in vivo
potential levels. However, it is felt that such
levels can be attained in vitro, using blood as
an electrolyte, collagen lined plumbing and
pulsatile blood flow conditions as occur in vivo.

In summary, it has been shown that pulsatile
flow of saline electrolytes generates electrokinetic potentials remarkably similar to in vivo
EKG's. This fact, in conjunction with prior
research(2), indicates that the present assumption
that EKG potentials are due solely to muscle
action potentials needs to be re-examined.

Acknowledgments

The assistance of Linda Stormo and Sidney
Golden of our research staff in the conduct of
tests and preparation of this paper is gratefully
acknowledged.

Abstract

Several left ventricle/aorta mechanical
simulators were fabricated to evaluate the possibility of generating EKG like electrical signals
by electrokinetic methodology. The simulators
produced pulsed turbulent flows, simulating
mammalian heart pumping conditions. EKG like
signals were generated by the motion of the
electrolyte through the simulators.

Literature Cited

1. Iberall, A., Cardon, S., Young, E., "On Pulsatile and Steady Arterial Flow, the GTS Contribution," General Technical Services Inc., Upper Darby, Pa. (1973) LC 72-96894.
2. Miller, J. R., Dent, R. F., Lab and Clin. Med., (1943), 28, 168.
3. Sawyer, P. N., Himmelfarb, E., Lustren, I., Ziskind, H., Biophys. J., (1966), 6, 641.
4. Cignette, M., "Streaming Potentials, Theory and Examples in Biological Systems," in Proc. 1st Inter. Symp. Biol. Aspects of Electrochem., S. Millazo, P. E. Jones, L. Rampazzo eds., Birkhauser Verlag, Basel, (1971).
5. Kupfer, E., J. Laryngology & Otology, (1938), 53, 16.
6. Srinivasan, S., Sawyer, P. N., J. Coll. Interfac. Sci., (1970), 32, 456.
7. Quinke, G., Ann. Physik, (1859), 2, (107), 1.
8. Zollner, F., Ann. Physik, (1873), 2, (148), 640.
9. Edlund, E., Ann. Physik, (1875), 2, (156), 251.
10. Haga, H., Ann. Physik, (1877), 3, (2), 326.
11. Clark, J. W., Ann. Physik, (1877), 3, (2), 335.
12. Helmholtz, H. L. F., Ann. Physik, (1879), 3, (7), 337.
13. Dorn, E., Ann. Physik, (1880), 3, (9), 513.
14. Boumans, A. A., Physica, (1957), 23, 1038.
15. Kurtz, F., Findl, E., Kurtz, A., Stormo, L., J. Coll. Interfac. Science, (in press).
16. Procopiu, S., Ann. Physik, (1913), 37, 229.
17. Zucker, E. R., "A Critical Evaluation of Streaming Potential Measurements," Ph.D. Thesis, Columbia Univ., (1959).
18. Newberry, A., Trans. Electrochem. Soc., (1934), 67, 25.
19. Ueda, T., et al., J. Electrochem. Soc. Japan, (1951), 19, 142.
20. Yeager, E., Hovorka, F., J. Acoustical Soc. America, (1953), 25, 445.
21. Williams, M., Rev. Sci. Instr., (1948), 19, 640.
22. Packard, R. G., J. Chem. Phys., (1953), 21, 303.

12

Differential Pulse Polarographic Analysis for Ethylenediaminetetraacetate (EDTA) and Nitrilotriacetate (NTA) in Phytoplankton Media

RICHARD J. STOLZBERG

Harold Edgerton Research Laboratory of the New England Aquarium, Central Wharf, Boston, Mass. 02110

In the course of investigating the effects of trace metal speciation on phytoplankton productivity, we have found it desirable to measure the concentration of small quantities of strong organic ligands. These include artificial ligands such as ethylenediaminetetraacetate (EDTA), nitrilotriacetate (NTA), and tris(hydroxymethyl)aminomethane (tris) added by the experimenter and some less well-defined extracellular metal binding organics (EMBO) added by the phytoplankton. Artificial ligands have traditionally been added by algal physiologists because a wide variety of algae can be grown in media containing complexed trace metals. In addition, precipitation of the medium is reduced, enabling the experimenter to prepare a more reproducible medium (1). The hypothesis has been made that plankton might actively produce EMBO for much the same reason - to improve the medium for growth either by detoxifying potentially toxic metals such as copper (2) or by making iron available as a soluble chelated species (3,4).

Our work currently involves correlating the concentration of complexed and uncomplexed species of copper with phytoplankton productivity and with the production of EMBO. Well defined artificial media that lend themselves to convenient chemical manipulation and which support a good growth of algae are used. Table I gives the composition of both the synthetic seawater (SSW) used in the analytical development work and the artificial medium designated Cu-IV.

In media containing less than 5×10^{-6} M EDTA or NTA, accurate measurement of specific ligand concentration is quite important. Small variations in ligand concentration may in certain cases produce

TABLE I

Composition of Synthetic Seawater and Cu-IV

Component	SSW	Cu-IV
NaCl	4.3×10^{-1} M	4.3×10^{-1} M
KCl	9.4×10^{-3} M	9.4×10^{-3} M
$MgSO_4$	2.7×10^{-2} M	3.7×10^{-2} M
$CaCl_2$	9.5×10^{-3} M	7.5×10^{-3} M
$NaNO_3$		1.2×10^{-3} M
NaH_2PO_4		4.8×10^{-5} M
Na_2SiO_3		2.6×10^{-4} M
Boron		4.4×10^{-5} M
EDTA		5×10^{-7} to 5×10^{-6} M
NTA		5×10^{-7} to 5×10^{-6} M
Tris		8×10^{-5} to 8×10^{-3} M
Mn		2.2×10^{-6} M
Zn		2.2×10^{-7} M
Co (inorganic)		5.4×10^{-8} M
Vitamin B_{12}		3.7×10^{-9} M

important changes in copper speciation. Loss of
ligand due to photodegradation is a distinct possibility when culturing algae, particularly in long
term experiments under high light intensity. The
sensitivity of ferric-EDTA to photodegradation has
been known for over two decades (5). The susceptibility of both ferric- and cupric-NTA to photodegradation has been documented recently (6,7,8).
Sorption of ligands can also present experimental
difficulties in dense algal cultures. Changes in
metal speciation can be expected from any of these
mechanisms that might reduce ligand or metal concentrations. Finally, the production of EMBO by the
cells should be taken into consideration when
calculating speciation. The quantity of EMBO produced
by the cells could be determined by the difference in
total complexing capacity (9) and the quantity of
EDTA or NTA present.

The analytical methodology for specific and
sensitive determination of EDTA and NTA in saline
waters is not well developed in spite of the wide
range of techniques developed for NTA in fresh water
and sewage sludge (10). The most widely used
specific techniques in non-saline water are gas
chromatography (11-13) and electrochemistry (14-18).
The presence of large quantities of dissolved salts
in seawater clearly favors electrochemical techniques.
Electrochemical reduction of the CdNTA complex,
first used analytically for the determination of NTA
in EDTA (14) and then later adapted for NTA determination in lake and river waters (16,17), is the
system of choice. Reduction of NTA complexes of
lead, bismuth, and indium (15,18) has been used
analytically, but the CdNTA (19,20) and CdEDTA (21,
22,23) electrochemistry has been characterized in
detail and appears relatively well behaved.

Classical DC polarography has been used to
measure 1 to 10 ppm NTA in lake water (16), but this
technique could not be used in seawater. The
reduction current due to the large quantity of cadmium
added to displace calcium from the NTA would swamp
the small current increments due to CdNTA reduction.
Differential pulse polarography (DPP) is more
sensitive than DC polarography, and it can be used
to measure small currents at a potential cathodic
of an electrochemically active species present at
much greater concentration.

Theoretical Considerations

The reduction of the CdNTA complex at -0.9 V (vs SCE) is irreversible and diffusion controlled (20). At pH 8 the reduction of the CdEDTA complex at -1.2 to -1.3 V is more irreversible than that for CdNTA. It is diffusion controlled only when the concentration of the support electrolyte is high (21). Separation of the uncomplexed Cd (at -0.6 V), the CdNTA, and the CdEDTA waves will present no problems using DPP. In seawater, the CdEDTA complex reduction current is expected to be diffusion controlled due to the presence of a high concentration of salts, and the sensitivity should be sufficient for the determination of micromolar quantities of ligand. The addition of a large excess of cadmium will cause a large fraction of both ligands to be present in the bulk of the solution as the cadmium complex (see below).

Equation 1 describes the generalized competition reaction between cadmium and competing metal (M) for EDTA or NTA (L). Charges have been omitted for clarity. Concentration stability constants and molar concentrations are used throughout.

$$CdL + M \rightleftharpoons Cd + ML, \quad K_C = \frac{K_{ML}}{K_{CdL}} \tag{1}$$

$$\text{where } K_{ML} = \frac{[ML]}{[M] \, [L]} \tag{1a}$$

$$\text{and } K_{CdL} = \frac{[CdL]}{[Cd] \, [L]} \tag{1b}$$

Using Ringbom's concept of conditional stability constants and his notation (24), the competition stability constant, K_C, can be rewritten as a conditional competition constant

$$K'_C = \frac{K'_{ML}}{K'_{CdL}} = \frac{[Cd'] \, [ML]}{[M'] \, [CdL]} \tag{2}$$

where M' refers to all of M not associated with L.

In the case where the cadmium concentration, C_{Cd}, is much greater than the EDTA or NTA concentration, C_L, the following approximation is valid:

$$[ML] \approx C_L - [CdL] \tag{3}$$

Substituting equation 3 into equation 2 and re-arranging results in

$$[CdL] = \frac{[Cd']}{K_c' [M'] + [Cd']} C_L \qquad (4)$$

Under the conditions of the analysis in seawater (where M = Ca), the following approximations are also valid:

$$[M'] \approx C_M \qquad (5)$$

$$[Cd'] \approx C_{Cd} \qquad (6)$$

Substituting into equation 4 produces the following relationship:

$$[CdL] = \frac{C_{Cd}}{(K_c' C_M) + C_{Cd}} C_L \qquad (7)$$

Equation 7 predicts that the concentration of electroactive CdL species in the bulk of solution is a linear function of the analytical concentration of L when C_M and C_{Cd} are constant. Under the conditions of analysis in synthetic seawater (2.4×10^{-4} M Cd), approximately 80% of the NTA and 100% of the EDTA are associated with cadmium. Figure 1 plots the fraction of NTA and EDTA associated with cadmium as a function of cadmium concentration.

A complication can exist if the method of standard additions is used. If enough ligand is added such that the condition $C_{Cd} \approx C_L$ no longer holds, equation 6 is no longer valid. However, the relationship

$$[Cd'] = C_{Cd} - [CdL] \qquad (8)$$

can be substituted into equation 4 and, after rearrangement, the following quadratic equation results:

$$[CdL]^2 - (K_c' C_M + C_{Cd} + C_L) [CdL] + C_{Cd} C_L = 0 \qquad (9)$$

The solution of this equation of [CdL] predicts a non-linear plot of current vs. concentration of ligand, the curvature of which is particularly pronounced at high concentrations of ligand. This effect will be discussed in a later section, specifically for NTA.

Experimental

   Apparatus. The electrochemical system used was
a Princeton Applied Research Model 174 Polarographic
Analyzer and a Model 172A mercury drop-timer. Output
was to a Houston Instruments Omnigraphic 2000 X-Y
recorder. The electrochemical cell was similar to
the one described by Gilbert and Hume (25), but a
dropping mercury electrode (DME) was used rather than
a wax impregnated graphite electrode. Synthetic
seawater was used in the seawater-silver reference
electrode. Mixing of the analyte was done with a
magnetic stirrer and Teflon coated stirring bar.
Data analysis was performed on a Wang 600 program-
mable calculator.

   Reagents. Glass distilled, deionized water;
reagent grade chemicals; and triple distilled mercury
were used throughout this study. Prepurified
nitrogen gas presaturated with water was used to
deoxygenate samples.

   Procedure. Optimum conditions for the determination of NTA and EDTA differed as shown in Table II.
With the smaller quantity of cadmium added, baseline
noise in the region of the CdEDTA wave was reduced.
However, the linear range was extended from 4×10^{-5}
M to 2.4×10^{-4} M EDTA if the cadmium spike was
increased from 0.1 ml to 0.6 ml.

TABLE II

Analytical Procedure[a]

	NTA	EDTA
500 ppm cadmium	0.60 ml	0.10 or 0.60 ml
Equilibrate cadmium spike	No	Yes, if C_{EDTA} $<10^{-5}$ M
Increase calcium by 0.1 M	No	Yes
Deoxygenate	10 min	10 min
Voltage scan	-0.7 to -1.1V	-1.0 to 01.6V
Peak potential[b]	-0.90V	-1.25V

(a) 10.00 ml sample + 1.00 ml tris   pH 7.9
(b) pulse amplitude 25 or 50 mV, scan rate 5 mV sec^{-1}

The method of standard additions was generally used in evaluating the technique. However, linear calibration curves could be made for NTA if the sum of strong ligands present was $<10^{-4}$ M and for EDTA if EDTA concentration was less than the cadmium concentration. Thus, for analysis of a wide variety of media, standard additions were not necessary. It was therefore possible to do simultaneous determinations of NTA and EDTA if the total strong ligand concentration was $<10^{-4}$ M. Standard additions could be made (keeping the total final ligand concentration below 10^{-4} M) or calibration curves could be used. In neither case was calcium added because the sensitivity of the NTA determination dropped precipitously. The quantity of cadmium added was 0.6 ml to ensure adequate sensitivity in the NTA determination.

Results and Discussion

General Evaluation. Figure 2 shows a differential pulse polarogram of the cadmium complexes of NTA and EDTA in SSW. With two minor exceptions, the theoretical considerations have been proven valid. One exception to the behavior predicted by theory is that the quantity of calcium present in SSW was actually not sufficient to produce a wave that is diffusion controlled. This deviation from theory does not seriously limit the technique because the wave height is nonetheless proportional to C_{EDTA}. The second exception is that simple competitive theory does not predict electrode response when copper competes with cadmium for the ligand.

Hydrogen ion control is necessary to keep pH in the range of 7 to 8. Above pH 8 the added cadmium tends to precipitate. Below pH 6 or 7, the rapid dissociation reaction of the protonated CdHNTA species becomes important (20).

$$CdNTA^- \rightleftharpoons Cd^{2+} + NTA^{3-} \quad Slow \qquad (10)$$

$$CdHNTA \rightleftharpoons Cd^{2+} + HNTA^{2-} \quad Fast \qquad (11)$$

The sensitivity of the technique is reduced in even slightly acid solution because the concentration of the species reduced at -0.9 V is diminished. Formation of the CdHEDTA$^-$ species at pH 5 plays a similar role in decreasing the reduction current at -1.25 V.

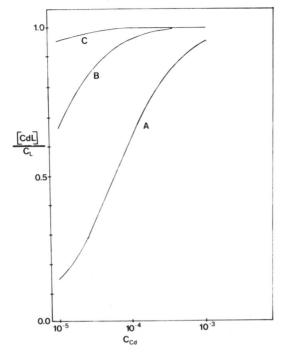

Figure 1.   Fraction of ligand associated with Cd in SSW as function of C_{Cd}.   (A) NTA, $K_c'C_M = 5.99 \times 10^{-5}$; (B) EDTA, 0.1 M Ca^{2+} added, $K_c'C_M = 5.06 \times 10^{-6}$; (C) EDTA, $K_c'C_M = 4.76 \times 10^{-7}$.

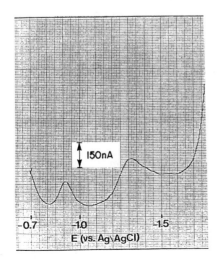

Figure 2.   Differential pulse polarogram of 1×10^{-5} M NTA and 2.5×10^{-5} M EDTA in SSW. 25-mV pulse height, 5-mV/sec scan rate, 1-sec drop time.

   Linearity. Within the limits of the accuracy of
the technique, the peak current at -0.9 V is a linear
function of NTA concentration from zero to 1 x 10^{-4} M
in SSW containing 2.67 x 10^{-4} M Cd. Solution of
equation 9 predicts that the fraction of NTA associated with cadmium decreases from 0.816 to 0.761 in
that range due to depletion of uncomplexed cadmium,
but this decrease is not apparent from the data. A
definite decrease in relative response is observed
experimentally above 10^{-4} M NTA. Figure 3 compares
experimentally observed behavior with that predicted
with no competition ($K_C' = 0$), and with two non-zero
values of K_C'. The value of K_C' calculated for calcium
competition using constants in reference 24 is $10^{-2.2}$.
The good fit of the experimentally observed points
to the line calculated for $K_C' = 10^{-2.2}$ suggests that
the system behaves as predicted when NTA is the
ligand.
   A plot of EDTA concentration in SSW vs. peak
current at approximately -1.25 V is linear from
10^{-5} M to 2.5 x 10^{-4} M in the presence of 2.67 x 10^{-4}
M Cd. Response at 5 x 10^{-6} M is generally less than
that expected by extrapolating the linear portion of
the calibration curve back toward the origin. This
decrease in relative response is due to the slow
reaction between micromolar quantities of EDTA and
cadmium. Maljkovic and Branica (26) observed that
the reaction between 2 x 10^{-6} M Cd and 2 to 7 x 10^{-6}
M EDTA in natural seawater proceeded only 31-69%
of the way to the equilibrium value in 10 min. In
the presence of 2.5 x 10^{-5} M EDTA the reaction is
rapid, however. In our usual analytical technique,
the EDTA present in the sample reacts with the
cadmium for only the 10 minute deoxygenation period.
We also have observed experimentally that the
reaction is incomplete in 10 minutes in spite of the
relatively high concentration of cadmium present.
When the sample is equilibrated with the cadmium
spike for an hour, the decreased relative response is
not observed with 5 μM EDTA present and linearity has
been observed to extend to this level and below.
   Using classical DC polarography, Raspor and
Branica (21) observed that the limiting current of
the CdEDTA wave is strongly dependent on the concentration, charge, and nature of the supporting
electrolyte cations. Addition of 10^{-1} M calcium to
dilute NaCl solutions containing EDTA increases
sensitivity by increasing relative response, as shown
in Table III. The addition of calcium to seawater

likewise increases the sensitivity of the technique for EDTA in that medium.

TABLE III

Effect of Ca^{2+} and Mg^{2+} on CdEDTA Peak Current [a]

Calcium or Magnesium Concentration (molar)	CdEDTA Current Ca^{2+} Additions [b] (nA)	CdEDTA Current Mg^{2+} Additions [b] (nA)
None	0.0	0.0
1×10^{-3}	8.0	2.7
5×10^{-3}	27.8	9.0
1×10^{-2}	34.1	15.1
5×10^{-2}	46.9	26.3
1×10^{-1}	51.0	30.00

(a) 5×10^{-5} M EDTA, 2.4×10^{-4} M Cd, 10^{-3} M NaCl, pH 7.9
(b) Mean of duplicate voltage scans

Precision and Accuracy. The precision of the analysis for NTA in SSW has been determined by replicate analysis of samples of NTA alone and in the presence of EDTA. Table IV presents results for five such experiments. The standard deviation of the analysis in the 1.5 to 5 µM NTA range is approximately 6×10^{-7} M. Accuracy is excellent except when simultaneous standard additions of NTA and EDTA are made to greater than 10^{-4} M total ligand.

Table V presents analogous results for EDTA determinations in SSW and in the medium Aquil (27), which resembles Cu-IV. The accuracy is excellent for 5×10^{-6} to 5×10^{-5} M EDTA. The precision increases significantly with the addition of 10^{-1} M Ca^{2+}, although accuracy is not affected. If the cadmium spike is allowed to equilibrate with the sample, satisfactory analyses can be made of samples containing less than 1×10^{-5} M EDTA.

The precision of the replicated data would suggest that the limit of detection in seawater is approximately 1 µM NTA and 1 µM EDTA. The detection limit for NTA indeed is approximately 1 µM.

TABLE IV

Precision and Accuracy of NTA Determination[a]

Sample	NTA Found[b]	Standard Deviation	Replicates
1.57 µM NTA in SSW	1.63 µM	0.68 µM	6
5.24 µM NTA in SSW	5.40 µM	0.68 µM	5
5.24 µM NTA in Cu-IV	5.50 µM	0.45 µM	6
5.24 µM NTA + 50 µM EDTA in SSW[c]	10.6 µM	0.69 µM	5
5.00 µM NTA + 50 µM EDTA in SSW[d]	4.77 µM	0.48 µM	6

(a) 25 mV pulse amplitude, 5 mV/sec scan rate, 0.6 ml 500 ppm Cd per 10 ml sample
(b) Standard addition analysis
(c) Simultaneous standard additions of NTA and EDTA. Final strong ligand concentration 2.2×10^{-4} M
(d) Additions of NTA only

As seen in Figure 4, the limiting factors are the small reduction current, the presence of baseline noise, and the rapidly decreasing current from the uncomplexed cadmium reduction. For EDTA, the detection limit is more nearly 2 or 3 µM. Figure 5 presents a polarogram of 5 µM EDTA in Aquil, and once again the small current and noisy baseline are important factors. Response does drop off sharply near 2 to 3 µM EDTA, and this phenomenon appears to set the lower limit of detection at this time.

Competitive Reactions. For this technique to be successful, a large and constant fraction of NTA and EDTA must be associated with cadmium near the electrode surface. In most polarographic analyses it is

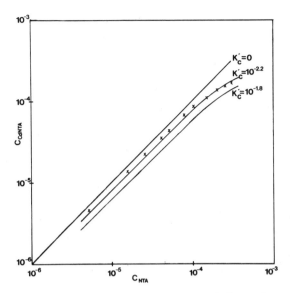

Figure 3.   Concentration of CdNTA in bulk of solution as function of total NTA concentration. Cadmium concentration = 2.67 × 10⁻⁴ M; × = experimental points calculated from wave height; — = calculated lines for various values of K_c'.

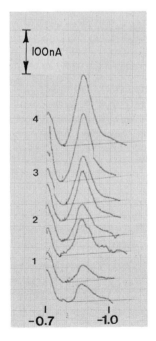

Figure 4.   Polarogram of 1.6 μM NTA and standard additions. 25-mV pulse, 5-mV/sec scan rate, 1-sec drop time. (1.) 1.6 μM NTA, duplicated; (2.–4.) spiked with successive 1.6-μM NTA increments, duplicated.

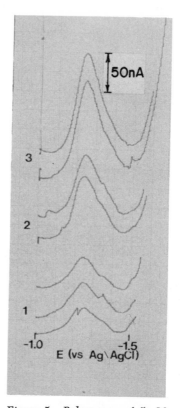

*Figure 5. Polarogram of 5 μM
EDTA and standard additions.
Same conditions as Figure 4.
(1.) 5 μM EDTA, in triplicate;
(2., 3.) spiked with successive 5-
μM EDTA increments, duplicated.*

TABLE V

Precision and Accuracy of EDTA Determination[a]

Sample	EDTA Found	Standard Deviation	Replicates
5.0×10^{-6} M EDTA in Aquil, 0.1 M Ca^{2+} added	3.0×10^{-6} M	0.8×10^{-6} M	5
5.0×10^{-6} M EDTA in Aquil, 0.1 M Ca^{2+} added[b]	4.48×10^{-6} M	0.42×10^{-6} M	4
1.00×10^{-5} M EDTA in SSW	0.93×10^{-5} M	3.5×10^{-6} M	6
5.00×10^{-5} M EDTA + 5.24×10^{-6} M NTA in SSW[c]	5.09×10^{-5} M	5.0×10^{-6} M	6
5.00×10^{-5} M EDTA in SSW	5.55×10^{-5} M	3.1×10^{-6} M	5
5.00×10^{-5} M EDTA in SSW, 0.1 M Ca^{2+} added	5.13×10^{-5} M	1.1×10^{-6} M	5

[a] 25 mV pulse amplitude, 5 mV/sec scan rate, 0.6 ml 500 ppm Cd added per 10 ml sample except where noted, standard addition analysis.

[b] 0.1 ml 500 ppm Cd added per 10 ml sample, equilibrate cadmium spike 1 hr before analysis.

[c] Simultaneous additions of NTA and EDTA.

desirable to know if the solution near the electrode
surface reflects the speciation and concentration of
electrochemically active components in the bulk of
solution. We have studied the competition between
cadmium and a number of other metals for both
ligands. When the competing metal is calcium, ferric
iron, zinc, nickel, or cobalt, the observed reduction
of the height of the CdL wave is proportional to the
reduction in the calculated fraction of L associated
with Cd. When the competing metal is copper, the
observed wave is much larger than is expected from a
calculation of ligand speciation in the bulk of
solution.

The copper competition experiments have demonstrated two points. A) The analytical technique works
well in the presence of moderate quantities of copper
although thermodynamics alone might predict that it
won't. B) Metal speciation measurements made in water
using electrochemical techniques can be misleading if
care is not taken in interpreting results. In the
system studied here the equilibrium established near
the electrode is quite different than that in the
bulk of solution due to an accumulation of ligand
near the electrode surface.

   Analysis of Media. The preparation of phytoplankton growth medium requires the addition of
nutrients (N, P, Si) and micronutrients (trace metals,
ligands, vitamins) to SSW. Results presented in
Tables IV and V demonstrate the applicability of the
technique to samples of medium as well as synthetic
seawater. No decrease in precision or accuracy was
observed in either Cu-IV or Aquil.

We have followed photodegradation of the artificial medium ASP-7 (8.1 x 10^{-5} M EDTA, 3.66 x 10^{-4} M
NTA) under rather extreme conditions (direct sun-
light, mid-summer, no temperature controls) and
observed rapid, but sometimes erratic disappearances
of both NTA and EDTA. No such decreases were observed
under normal culturing conditions over a period of
two weeks. The relative standard deviation of
replicate analyses was $\leq 5\%$.

The presence of even moderate densities of
Skeletonema costatum, a marine diatom, in the medium
decreases the reliability of measurements. Centri-
fugation of the sample is thus necessary for analyzing
growing cultures.

This technique is useful for measurement of NTA
and EDTA concentrations in a wide variety of media.
If degradation or sorption processes are occurring

and metal speciation is of interest, a specific technique such as this one is necessary. This is particularly so if both ligands are present, as in ASP 7. The precision of the technique limits its ability to measure submicromolar changes in ligand concentration. If degradation is not occurring, a non-specific technique could be used (9). However, this technique is the method of choice if specific measurements of NTA and EDTA in seawater are needed at the micromolar level.

Abstract

A technique for the analysis of synthetic seawater and phytoplankton media for ethylenediaminetetraacetate (EDTA) and nitrilotriacetate (NTA) by differential pulse polarography has been developed. The addition of approximately 2.4×10^{-4} M cadmium to the analyte converts a large fraction of either ligand to the reducible cadmium complex. With 5×10^{-6} M ligand present, the standard deviation of the technique is approximately 6×10^{-7} M for NTA and 4×10^{-7} M for EDTA. An examination of how accurately the electrode reaction reflects the speciation of these ligands in the bulk solution has also been made. With calcium, nickel, cobalt, and zinc added as competing cations, the electrode response is as expected. In the presence of copper, the proportion of ligand associated with cadmium, as indicated by the reduction current, was much greater than predicted by thermodynamic considerations alone.

Credit

This research was supported by the Oceanography Section, National Science Foundation, Grant DES74-21642.

Literature Cited

1) Provasoli, L., McLaughlin, J. J. A., and Droop, M. R., Archiv. für Mikrobiologie (1957), 25, 392.
2) Steemann Nielsen, E. and Wium-Andersen, S., Physiol. Plant. (1971), 24, 480.
3) Murphy, T. P., Lean, D. R. S., and Nalawajko, C., Science (1976), 192, 900.
4) Levandowsky, M. and Hutner, S. H., Ann. N.Y. Acad. Sci. (1974), 245, 16.
5) Jones, S. S. and Long, F., J. Phys. Chem. (1952), 56, 25.

6) Stolzberg, R. J. and Hume, D. N., Environ. Sci. Technol. (1975), 9, 654.

7) Langford, C., Wingham, M., and Sastri, V., ibid. (1973), 7, 820.

8) Trott, T., Henwood, R., and Langford, C., ibid. (1972), 6, 367.

9) Stolzberg, R. J. and Rosin, D., "Complexing capacity measurements using the sodium form of Chelex 100: General theory and application to seawater-like matrices", presented at 172nd National ACS Meeting, San Francisco, CA (1976).

10) Mottola, H. A., Toxicol. Environ. Chem. Rev. (1974), 2, 99.

11) Stolzberg, R. J. and Hume, D. N., Anal. Lett. (1973), 6, 829.

12) Aue, W., Hastings, C., Gerhard, K., Pierce, J., Hill, H., and Moseman, R., J. Chromatogr. (1972), 72, 259.

13) Warren, C. and Malec, E., ibid. (1972), 64, 219.

14) Daniel, R. and LeBlanc, R., Anal. Chem. (1959), 31, 1221.

15) Afghan, B. and Goulden, P., Environ. Sci. Technol. (1971), 5, 601.

16) Asplund, J. and Wanninen, E., Anal. Lett. (1971), 4, 267.

17) Wernet, J. and Wahl, K., Fresenius' Z. Anal. Chem. (1970), 251, 373.

18) Haberman, J. P., Anal. Chem. (1971), 43, 63.

19) Koryta, J. and Kossler, I., Collect. Czech. Chem. Commun. (1950), 15, 241.

20) Raspor, B. and Branica, M., J. Electroanal. Chem. (1975), 59, 99.

21) Raspor, B. and Branica, M., ibid. (1973), 45, 79.

22) Raspor, B. and Branica, M., ibid. (1975), 60, 35.

23) Schmid, R. W. and Reilley, C. N., J. Amer. Chem. Soc. (1958), 80, 2101.

24) Ringbom, A., "Complexation in Analytical Chemistry", p. 35, Interscience, N.Y. (1963).

25) Gilbert, T. R. and Hume, D. N., Anal. Chim. Acta (1973), 65, 451.

26) Maljkovic, D. and Branica, M., Limnol. Oceanogr. (1971), 16, 779.

27) Morel, F. M. M., Westall, J. C., Reuter, J. G. and Chaplick, J. P., Technical Note #16, Ralph M. Parsons Lab., M.I.T., Cambridge, MA, Sept. 1974.

INDEX